GADOURI Hamid
MEZIANI Brahim

Road pavement deterioration, road testing and formulation

GADOURI Hamid
MEZIANI Brahim

Road pavement deterioration, road testing and formulation

Expertise in road pavement condition surveys and testing of bituminous materials and concretes

ScienciaScripts

This book is a translation from the original published under ISBN 978-620-3-45898-5.

Publisher:
Sciencia Scripts
is a trademark of
Dodo Books Indian Ocean Ltd. and OmniScriptum S.R.L publishing group

120 High Road, East Finchley, London, N2 9ED, United Kingdom
Str. Armeneasca 28/1, office 1, Chisinau MD-2012, Republic of Moldova, Europe
Printed at: see last page
ISBN: 978-620-7-30148-5

Contents

Foreword

This book, entitled **"DETERIORATION OF ROAD PAVEMENTS, ROAD TESTING, AND FORMULATION"**, is aimed at students of engineering geology and geotechnics, particularly those specialising in road engineering. It provides a summary of the basic theoretical and practical knowledge relating to road pavement construction, focusing on three key chapters.

The first chapter, entitled "Criteria, methods and procedures for assessing road pavement damage", discusses the various criteria and methods used to assess pavement damage.

The second chapter, entitled "Road Identification Tests", focuses on the tests used to identify the characteristics and properties of road pavements. It examines the various test methods, such as laboratory and in-situ tests, for assessing the performance of the materials used in pavement construction, as well as aspects of traffic loads and environmental conditions in pavement design.

The third chapter, entitled "Formulation of bituminous mixes", looks at the formulation and properties of bituminous mixes used in pavement construction, exploring the components of bituminous mixes, the mixing and manufacturing processes, as well as aspects relating to the durability and performance of pavements.

Overall, this book provides readers with an in-depth understanding of the various stages in the construction and re-evaluation of road pavements. It conforms to the official syllabus of the Algerian Ministry of Higher Education and Scientific Research (MESRS) and aims to help future engineers develop the skills needed to meet the challenges of road pavement structures in their professional practice.

GENERAL INTRODUCTION

The book **"Degradation of road pavements, road testing and formulation"** offers a comprehensive and in-depth approach to the construction and re-evaluation of road pavements. It is aimed at students of engineering geology and geotechnics, focusing on the essential aspects of road engineering.

The chapters in this book are carefully structured to cover the main areas related to road pavement structures. The first chapter, **"Criteria, mode and methods of assessing road pavement damage"**, discusses the criteria and methods used to assess pavement damage.

The second chapter, **"Road Identification Tests"**, focuses on the tests used to identify the characteristics and properties of road pavements, with emphasis on laboratory and field (in-situ) testing methods. Traffic loads and environmental conditions are also examined in the context of pavement design.

The third chapter, **"Asphalt Concrete Formulation"**, looks specifically at the formulation and properties of asphalt concrete used in pavement construction, exploring components, mixing and manufacturing processes, as well as pavement durability and performance considerations.

This book has been developed in accordance with the official programme of the Algerian Ministry of Higher Education and Scientific Research (MESRS). Its main objective is to help future engineers acquire the skills needed to meet the challenges of road pavement structures in their professional practice.

In short, this book is an indispensable tool for students of road engineering, providing them with an in-depth understanding of the various stages in the construction, evaluation and formulation of asphalt concrete for road pavements. It aims to foster the development of sound skills and prepare future engineers to meet the practical challenges of the field.

CHAPTER I

CRITERIA, MODE AND METHODS OF EVALUATION OF DAMAGE TO ROAD PAVEMENTS

1.1 Introduction

The deterioration and wear of road surfaces are the result of several phenomena that occur during the life of the pavement. There are many causes and the modes of deterioration vary according to these causes. Mechanical and thermal stresses are the main causes of road surface degradation and must be taken into account during the pavement design phase and when selecting pavement components. The construction phase also has a major impact on the durability of the pavement.

Proper compaction of granular or bound layers, bonding of asphalt layers and proper execution of joints are important conditions for a durable road offering comfort and safety to users. Particular attention must therefore be paid to all these phenomena during the pavement design and construction phases. Understanding these modes of deterioration and wear is also important during the life of the pavement in order to ensure proper maintenance of the pavement and extend its life. The choice of maintenance techniques must imperatively involve a proper analysis of the causes of the deterioration observed. This subject could be the subject of another article in the future.

1.2 Criteria and method for assessing road damage

1.2.1   Evaluation criteria

The assessment of pavements is based on a series of measurements and visual observations that make it possible to establish the condition of the structure, diagnose the causes of apparent deterioration and target the most appropriate rehabilitation solutions. In the case of measurements such as the geometric or physical characteristics of the pavement, it is easier to establish criteria on which to base assessment and rehabilitation. In the case of observations aimed at characterising surface degradation and pavement condition, it becomes more difficult to establish such criteria (MTQ, AIMQ 2002).

In order to overcome this difficulty, it is very necessary to formalise the characterisation of surface defects on pavements and to produce a summary that could be accessible to the personnel involved in this activity: this summary, based in particular on a series of photos and sketches, makes it possible to categorise surface degradations on flexible pavements and to obtain a way of measuring their extent and severity in an objective, coherent way that is harmonised with the most common current procedures. The aim is to improve communication and facilitate comparisons by standardising the names and types of measurements of the impairments. Each type of deterioration is described in general terms and is the subject of a sheet listing the most likely causes of the deterioration and describing the three severity levels (low, medium and major). In general, for each deterioration, the three severity levels include the following concepts:

Low: This level corresponds to the initial stage of deterioration: the first signs sometimes appear intermittently on a segment of road and the assessor must be attentive to detect the symptoms of deterioration. This state is often difficult to perceive for an observer travelling in a vehicle at a speed of around 50 km/h. At the maximum speed permitted, ride comfort is not affected, or only to a very limited extent.

Medium : This summer indicates a continuous and easily perceptible deterioration for an observer travelling at a speed of around 50 km/h. At the maximum permitted speed, ride comfort is significantly reduced by most of the deterioration.

Major: This condition indicates that the degradation is accentuated and evident, even to an observer travelling at the maximum permitted speed. Ride comfort is generally reduced and, in some cases, safety at maximum speed may be compromised. A repair or correction operation should be considered as soon as possible when this condition is reached.

1.2.2   Assessment method

The AASHTO standard PP44-01 *"Quantifying Cracks in Asphalt Pavement Surface"* aims to simplify the classification of cracks and thus ensure more consistent measurement results over time and greater reliability for pavement management purposes. Part of the cracking block is based on this standard. The concept consists of dividing each lane surveyed into "5" strips, i.e. "2" strips for wheel tracks and "3" strips for non-wheel tracks. Taking into account the direction of traffic, strip number "1" is on the left-hand side and strip number "5" is on the right-hand side of the lane being tested (Figure I.1).

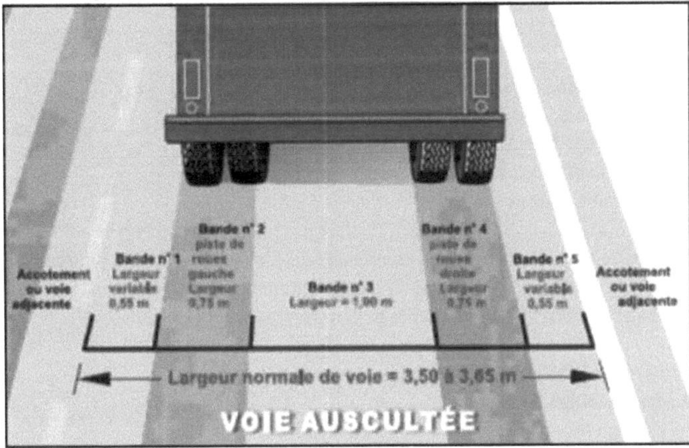

Figure I.1 - Details of monitoring strips for cracking (degradation) (MTQ-AIMQ 2002).

On this basis, a cracking rate for each strip can be established from the cracks detected in length and width. Cracks in wheel tracks are mainly associated with traffic loads and fatigue damage, whereas cracks in other strips are attributable to more varied causes, such as frost and ageing, or to construction defects. It should be noted, however, that often in urban environments or in the case of particular geometries with wide lanes, the principle of lane inspection strips for cracking may prove difficult to apply and even inappropriate.

1.3 Degradation of flexible and semi-rigid pavements

Generally speaking, the degradations observed in flexible and semi-rigid pavements can be classified into five main families, which are presented with photographic examples as follows (CTTP 1995, 1996; LCPC-IFSTTAR 1998; MTQ-AIMQ 2002):

Family of deformations ;

Cracking family ;

Tear-off family ;

Material movement family ;

Urban damage family.

1.3.1 Deformation family

These are degradations that lead to changes in the pavement, giving the surface a different appearance from that desired. These deformations, which originate in the body of the pavement, generally affect the lower layers before reaching the surface course and can be distinguished according to their shape or location as follows:

1.3.1.1 Small radius rut

The small radius rut is a single, double and sometimes triple longitudinal depression, around 250mm wide, located in the wheel tracks. The transverse profile of these depressions is often similar to single or twin tyre tracks.

1.3.1.1.1 Size and scope

The three levels of severity of a small radius rut are shown in Figure I.2. The extent represents the % of the total length of the affected areas in relation to the total length of the survey section.

Low Medium   Major

Figure I.2 - Gravity of a small-radius rut (MTQ-AIMQ 2002).

Low: the depth of the rut is less than 10 mm.

Medium: at this level, the rut is 10 to 20 mm deep.

Major: the depth of the rut is greater than 20 mm.

1.3.1.1.2   Probable causes

The granular materials that make up the base of flexible pavements are sometimes not very rigid. Repeated high vertical stresses cause plastic deformations which are reflected in permanent deformations on the surface of the pavement. The probable causes of rutting at small radii are :

Asphalt with reduced stability in hot weather (e.g.: bitumen too soft or overdosed) ;

Asphalt too weak to withstand heavy traffic;

Insufficient compaction of the asphalt during installation;

Wear of the surface coating (abrasion).

1.3.1.2 Large radius rut

It is characterised by a single longitudinal depression located in the wheel tracks. The transverse shape of the depression corresponds to that of a very steep parabolic curve.

1.3.1.2.1   Size and scope

The three levels of severity of a large radius rut are shown in Figure I.3. The extent represents the % of the total length of the affected areas in relation to the total length of the survey section.

Low Medium Major

Figure I.3 - Gravity of a large-radius rut (MTQ-AIMQ 2002).

Weak: the depth of the rut is less than 10 mm;

Medium: the depth of the rut is 10 to 20 mm;

Major: the depth of the rut is greater than 20 mm.

1.3.1.2.2   Probable causes

The most plausible and frequent causes of large radius ruts are as follows:

Ageing (accumulation of permanent deformations) ;

Insufficient compaction in granular layers during construction;

Insufficient structural capacity of the carriageway ;

Poor drainage of granular pavement materials (e.g. during thaw periods) and wear.

1.3.1.3 Subsidence

This is a distortion of the profile at the edge of the carriageway or in the vicinity of underground pipes, or a very pronounced and often fairly extensive depression located either on the edge or at full width.

1.3.1.3.1   Size and scope

The three levels of severity of subsidence are shown in Figure I.4. The extent is the % of the total surface affected in relation to the surface of the section surveyed.

Low Medium Major

Figure I.4 - Gravity of a subsidence (By S. Sihem 2017, Algeria).

Low: this is defined by a denivellation whose depth is less than 20 mm below the 3 m rule;

Medium: here the depth of cleavage is between 20 and 40 mm below the 3m rule; **Major: this** corresponds to a depth of cleavage greater than 40 mm below the 3m rule.

1.3.1.3.2 Probable causes

Collapses of flexible carriageways are often caused by the instability of the backfill, the presence of inadequate or poorly compacted materials, clayey overburden, drying out of the supporting soil, and the poor condition of underground networks (in urban areas). There are other reasons, such as local undersizing, pollution of the road surface or locally defective construction.

1.3.1.4 Differential lifting

Differential heave is defined as the localized swelling of the pavement during a period of frost, both parallel and perpendicular to the axis of the pavement.

1.3.1.4.1 Size and scope

The three levels of severity of differential heave are shown in Figure I.5. The extent is the total % of the area affected by this type of degradation in relation to the total surface area of the survey section.

Low: Progressive denivellation with a height of less than 50 mm;

Medium : Progressive denivellation with a height of between 50 and 100 mm ;

Major: Progressive denivellation with a height greater than 100 mm or sudden denivellation of any height.

Low Medium Major

Figure I.5- Gravity of a differential heave (MTQ-AIMQ 2002).

1.3.1.4.2 Probable causes

The most plausible causes are :

Frost infrastructure, a recurring winter phenomenon;

Moisture-sensitive materials, a permanent phenomenon ;

High water table and presence of water in the vicinity of the causeway;

Heterogeneity of materials or inadequate transition in the pavement ;

Shallow underground pipes (urban environment).

1.3.1.5 Profile disorder

The profile disorder is observed in the case of inappropriate slopes and geometry favouring the accumulation of run-off water in puddles on the surface of the carriageway.

1.3.1.5.1 Size and scope

The three levels of severity of a profile disorder are shown in Figure I.6. The extent is the % of the total surface area affected by this type of degradation in relation to the total surface area of the section surveyed.

Low Medium Major

Figure I. 6 - Gravity of a profile disorder (MTQ-AIMQ 2002).

Low: Accumulation of water to a depth of less than 20 mm ;

Means : Water accumulates to a depth of 20 to 40 mm ;

Major: Accumulation of water to a depth of more than 40 mm.

1.3.1.5.2 Probable causes

The phenomena that cause profile damage are generally undrained low points and subsidence along kerbs.

1.3.2 Cracking family

Repeated alternating bending stresses in the bituminous cover of a flexible pavement lead to fatigue degradation, in the form of cracks which are initially isolated and then gradually develop into a small-mesh crack. Cracking is defined as pavement failure along a line with or without breakage of the pavement body. They may affect the wearing course alone, or part or all of the pavement body.

1.3.2.1 Transverse cracks

Transverse cracks occur when the pavement breaks relatively perpendicular to the direction of the road, generally across the entire width of the carriageway.

1.3.2.1.1 Size and scope

The three levels of severity of transverse cracks are shown in Figure I.7. The extent is the % of total surface area of the affected zone in relation to the cross-sectional area of the survey.

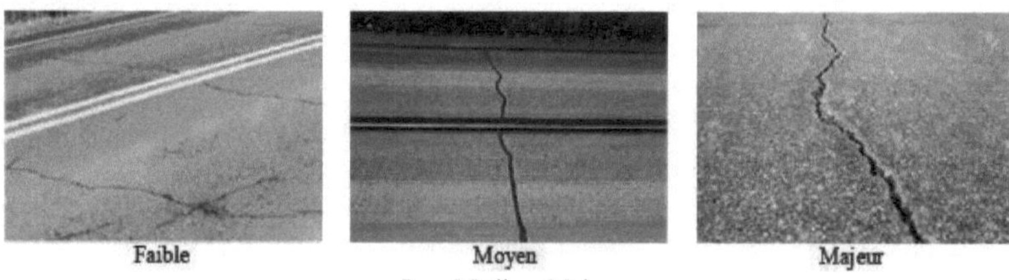

Low Medium Major

Figure I.7 - Gravity of transverse cracks (MTQ-AIMQ 2002).

Low: Simple, intermittent cracks with openings of less than 5 mm. The edges are generally clean and well defined;

Medium: Single cracks or multiple cracks along a main crack, the latter being open by 5 to 20 mm. The edges are sometimes eroded and slightly depressed. Without being uncomfortable, the crack is perceptible to the user;

Major: Single cracks or multiple cracks along a main crack, the latter being open by more than 20 mm. The edges are often eroded and there is subsidence or frost heave in the vicinity of the crack. Rolling comfort is reduced by surface deformation.

1.3.2.1.2 Probable causes

Transverse cracks are caused by the phenomena shown opposite:

Heat shrinkage ;

Ageing and embrittlement of bitumen ;

Poorly executed construction joint (asphalt laying work stopped and resumed);

Reduction in the cross-section of the covering.

Occasionally, pavement slides occur, which are very large movements of the surface layer due to both insufficient bonding with the base layer and insufficient pavement stability. They are initiated by diagonal cracking at the wheel arch and by clear parabolic cracks.

As the water infiltrates more easily, the phenomena accelerate: spalling at the edges of the cracks, with the material escaping, followed by the formation of potholes. If the pavement is left unmaintained, it will very quickly progress to complete destruction.

1.3.2.2 Longitudinal cracks in wheel tracks

They are characterised by a break in the surface parallel to the direction of the road and located in the wheel tracks.

1.3.2.2.1 Size and scope

The three levels of severity of longitudinal cracks in wheel tracks are shown in Figure I.8. The extent is the percentage of the length affected in relation to the section length of the survey.

Low Medium Major

Figure I.8 - *Gravity of longitudinal cracks in wheel tracks (MTQ-AIMQ 2002).*

Low: This low level is defined by simple, intermittent cracks with openings of less than 5 mm. The edges are generally clean and well defined;

Medium level: The medium level is characterised by single cracks or multiple cracks along a main crack, the latter being open by 5 to 20 mm. The edges are sometimes eroded and a little depressed. Without being uncomfortable, the crack is perceptible to the user;

Major: Major cracks are single cracks or multiple cracks along a main crack, which is open by more than 20 mm. The edges are often eroded and there is subsidence or frost heave in the vicinity of the crack. Tile cracks are also present.

1.3.2.2.2 Probable causes

These types of cracks are often caused by :

Pavement fatigue (heavy traffic) ;

Insufficient structural capacity of the carriageway.

Poor drainage of the granular layers of the pavement (e.g. during thawing).

In addition, variations in temperature at the surface of a pavement cause traction and contraction phenomena that lead to cracks. Heat, which softens the surface layers, accelerates the ageing of hydrocarbon products.

The heat-cold cycle alters the stability of materials, making bituminous surfaces brittle and therefore prone to cracking and spalling.

1.3.2.3 Longitudinal cracks outside the wheel track

They are defined by the breakage of the tarmac relatively parallel to the direction of the road, outside the wheel tracks.

1.3.2.3.1 Size and scope

The three levels of severity of longitudinal cracks outside the wheel track are shown in Figure I.9. The extent is the % of the length affected in relation to the section length of the survey.

Low Medium Major
Figure I.9 - *Gravity of longitudinal cracks outside wheel tracks (MTQ-AIMQ 2002).*

Weak: This is defined by simple, intermittent cracks with openings of less than 5 mm. The edges are generally clean and well defined;

Medium: This medium level shows single cracks or multiple cracks along a main crack, the latter being open by 5 to 20 mm. The edges are sometimes eroded and slightly depressed;

Major: Single cracks or multiple cracks along a main crack, the latter being open by more than 20 mm. The edges are often eroded and there is subsidence or frost heave in the vicinity of the crack.

1.3.2.3.2 Probable causes

Longitudinal cracks outside the wheel track are caused by the phenomena shown opposite:

Poorly executed construction joint along the adjacent track;

Segregation of the mix during laying (e.g.: centre of spreader) ;

Ageing of the coating ;

Advanced fatigue of the carriageway or undersizing of one or more layers;

Reduced bearing capacity of the supporting soil (poor drainage, waterproofing defects);

Poor operation of the structure (detached layers);

Poor quality of certain materials.

1.3.2.4 Frost cracks

They correspond to the breakage of the surfacing generating an active crack under the effect of frost, either rectilinear and located in the centre of the track or carriageway, or of lezardous appearance with no precise location on the carriageway.

1.3.2.4.1 Size and scope

The three levels of severity of frost cracks are shown in Figure I.10. The extent is the % of the length affected in relation to the section length of the survey.

Low Medium Major
Figure I.10 - *Gravity of frost cracks (MTQ-AIMQ 2002).*

Low: This level reflects simple, intermittent cracks with openings of less than 10 mm. The edges are generally clean and well defined;

Medium: Characterised by single cracks or multiple cracks along a main crack, the latter being 10 to 25 mm open. The edges are sometimes eroded and slightly depressed. Without being uncomfortable, the crack is perceptible to the user;

Major: These are generally single cracks or multiple cracks along a main crack, the latter being open by more than 25 mm. The edges are often eroded and there is subsidence or frost heave in the vicinity of the crack. Rolling comfort is reduced by surface deformation.

1.3.2.4.2 Probable causes

Frost cracks are caused by the following phenomena:

Infrastructure gelive and differential lifts;

Differential freezing behaviour ;

Unstable fill ;

Inadequate drainage

1.3.2.5 Tile cracks" restoration

Faienqage" cracks in tiles are represented by the breakage of the coating over more or less extensive areas, forming a cracking pattern with small polygonal meshes whose average size is of the order of 300 mm or less.

1.3.2.5.1 Size and scope

The three levels of severity of cracks in *"Faiengage"* tiles are shown in Figure I.11. The extent is defined by the percentage of the sum of the surfaces of the damaged areas in relation to the total surface of the section of the survey.

Weak: this is a mesh of simple cracks with clean edges;

Medium : Mesh composed of simple cracks with slightly deteriorated edges;

Major: Mesh composed of single cracks with deteriorated edges.

Low Medium Major

Figure I.11 *- Gravity of cracks in "Faiengage" tiling (MTQ-AIMQ 2002).*

1.3.2.5.2 Probable causes

The most common causes of cracks in tiles are :

Fatigue (e.g. insufficient coating thickness) ;

Ageing of the pavement (oxidation and embrittlement of the bitumen in the mix);

Insufficient load-bearing capacity.

1.3.2.6 Edge cracks

Edge cracks correspond to breaks in straight lines or arcs along the shoulder or kerb, or detachment of the pavement along the kerb.

1.3.2.6.1 Size and scope

The three levels of severity of edge cracks are shown in Figure I.12. The extent is the % of the total length affected by degradation in relation to the total length of the section of the survey.

Figure I.12 - Gravity of shoreline cracks (MTQ-AIMQ 2002).

Weak: This is defined by simple, intermittent cracks with openings of less than 5 mm. The edges are generally clean and well defined;

Medium: Single cracks or multiple cracks along a main crack, the latter being open by 5 to 20 mm. The edges are sometimes eroded and slightly depressed;

Major: Single cracks or multiple cracks along a main crack, the latter being open by more than 20 mm. The edges are often eroded and there is subsidence or frost heave in the vicinity of the crack.

1.3.2.6.2 Probable causes

The cracks at the edges are due to the phenomena shown opposite:

Lack of lateral support (e.g. narrow shoulders and steep slopes);

Discontinuity in the structure (e.g. widening) ;

Lateral input of run-off water into the pavement structure (urban environment); Draining of the supporting soil.

1.3.3 Tear-off family

The family of pull-outs only concerns damage affecting the wearing course in general.

1.3.3.1 Desenrobage

Stripping is the erosion and loss of large aggregates on the surface, producing a progressive deterioration of the pavement.

1.3.3.1.1 Size and scope

The three levels of severity of stripping are shown in Figure I.13. The extent is the % of the area affected in relation to the cross-sectional area of the survey.

Figure I.13 - Overburden gradient (MTQ-AIMQ 2002).

Low: this is a barely observable loss of putty or coarse aggregate, mainly in wheel tracks;

Medium: Easily observable loss of sealant, leaving large aggregates very visible, or loss of large aggregates, leaving a regular pattern of small cavities all over the surface;

Major: this is defined by a completely eroded surface and accentuated degradation in the wheel tracks (incipient rutting due to wear).

1.3.3.1.2 Probable causes

Stripping is caused by the following phenomena: wear and tear due to heavy traffic, under-dosed bitumen, use of hydrophilic aggregates, insufficient compaction, overheating or ageing of the asphalt (oxidation and embrittlement), increased stress in bend and braking zones (urban environment), insufficient binder-aggregate adhesion, laying in adverse weather conditions, and water stagnation on the pavement.

1.3.3.2 Pelade

This is the tearing away of the surface course in patches.

1.3.3.2.1 Size and scope

Figure I.14 shows the three levels of severity of *"slab tearing"*. The extent is the % of the surface affected in relation to the cross-sectional area of the survey.

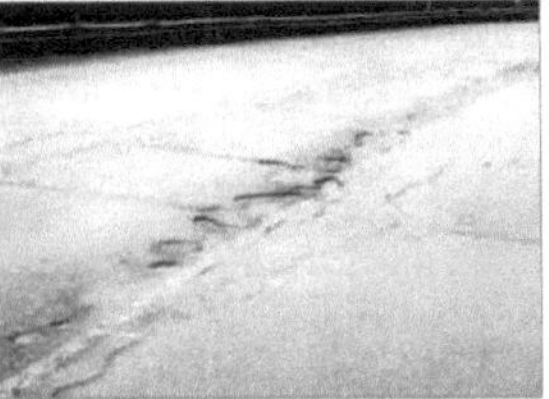

LowMedium   Major

Figure I.14 - Gravite of peladic tearing (MTQ-AIMQ 2002).

Low: Pelade with a pull-out surface of less than 0.5 square metres;
Medium: Peladic eruption with a tear surface of 0.5 to 1 square metre; **Major:** Peladic eruption with a tear surface greater than 1 square metre.

1.3.3.2.2 Probable causes

Hair loss is caused by the following phenomena:

Poor adhesion of the surface layer due to lack of bonding binder, chemical incompatibility and/or dirt between layers;

Insufficient thickness of the surface layer ;

Pavement subjected to heavy traffic.

1.3.3.3 Pothole

Potholes are the final manifestation of a combination of different problems. They are characterised by localized disaggregation of the pavement over its entire thickness, forming holes of varying size and depth that are generally rounded in shape and well-defined in outline.

1.3.3.3.1 Size and scope

The three levels of potholes are shown in Figure I.15. The extent is measured by the number of potholes per section of the survey.

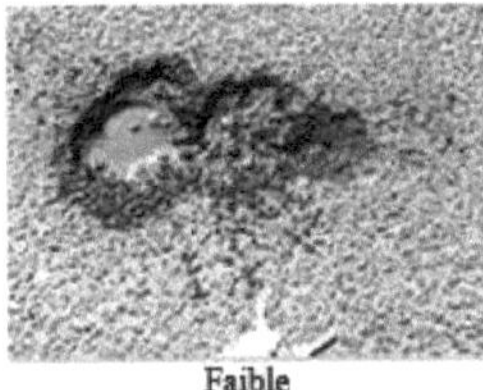

LowMedium   Major

Figure I.15 - Pothole gradient (MTQ-AIMQ 2002).

Low: this is described by a pothole with a diameter of less than 200 mm;

Medium: in this case the pothole has a diameter of 200 to 300 mm;

Major: at this level, the pothole is more than 300 mm in diameter.

1.3.3.3.2 Probable causes

Potholes are caused by local weaknesses in the sub-base, insufficient pavement thickness or poor bearing capacity (drainage, clay pockets, etc.). They can also occur when the carriageway is heavily loaded by heavy traffic, in the case of a local defect in the surface or base course, which is often due to poor quality in the manufacture or laying of the materials.

1.3.4 Movement of materials

1.3.4.1 Penetrant testing

Bleeding, by definition, is the rising of bitumen on the surface of the pavement, accentuated in wheel tracks.

1.3.4.1.1 Size and scope

The three levels of penetrant severity are shown in Figure I.16. The extent is the cumulative % length of the areas affected by the disorder in relation to the total length of the survey section.

Low: in this case, bleeding is mainly detectable in the wheel tracks by the appearance of a darker strip of coating;

Medium: here the wheel tracks are clearly delineated by the black tarmac;

Major: this corresponds to a wet, shiny appearance over most of the surface. The texture of the asphalt is impossible to discern. Tyre noise is similar to that produced on wet pavement. Most of the surface is affected.

LowMedium   Major

Figure I.16 *- Penetrant gravel (MTQ-AIMQ 2002).*

1.3.4.1.2   Probable causes

Bleeding is caused by overdosage of bitumen, the combined effect of high pavement temperature and traffic stresses, excess tack binder, mix design unsuited to stresses.

1.3.4.2 Glazing or "Indentation

Glazing, by definition, is the wearing away or sinking of chippings into the asphalt in hot weather under the action of traffic, giving the road surface a smooth, shiny appearance.

1.3.4.2.1   Size and scope

The three levels of ice severity or *"Indentation"* are shown in Figure I.17. The extent is the cumulative % length of the areas affected by the disorder in relation to the total length of the survey section.

LowMedium   Major

Figure I.17 *- Glazing gravel or "Indentation" (LCPC, IFSTTAR 1998).*

Low: perceptible, clear but isolated (of the order of a metre);

Medium: significant and concerns the treads in zones of around 10m;

Major: generalized phenomenon across the entire cross-section or in both treads in areas greater than 10m.

1.3.4.2.2   Probable causes

The causes of icing are :

Insufficient hardness of paving aggregates;

Overdosage of binder in the mix ;

The quality of the binder is unsuited to the traffic or climate;

Local repairs;

Site incidents;

Oil leaks from site machinery.

1.3.4.3 Rise of the fines

Appearance of fine elements on the surface of the pavement originating from the sub-base which are generally located in line with defects in the wearing course (cracks, pits, cracks, etc.).

I.3.4.3.1 Gravity and extent

The three levels of severity of the upwelling of fines are shown in Figure I.18. The extent is the cumulative % surface area of the zones affected by the disorder in relation to the total surface area of the section surveyed.

LowMedium   Major

Figure I.18 - *Upwelling gravity of fines (LCPC, IFSTTAR 1998).*

Low: perceptible and local degradation;

Medium: clear, significant and widespread degradation and more severe (fewer fines);

Major: frank, significant, extensive and more severe degradation (too many fines).

I.3.4.3.2 Probable causes

The rising of fines to the surface is generally caused by water circulating in the sub-base under the effect of pumping generated by traffic. These particles come from the subgrade soil (in the case of water-sensitive fine soils). This phenomenon can occur on cracks that are fluttering or on areas where the base has a laminate, surface decohesion or decohesion in the mass.

I.3.5 Damage in urban areas

1.3.5.1 Cracking around manholes and sumps

It is described by the breaking of the coating in a circular and/or radial pattern.

1.3.5.1.1   Size and scope

The three levels of severity of cracking around manholes and sumps are shown in Figure I.19. The extent is the total number of manholes per section of the survey and per severity level.

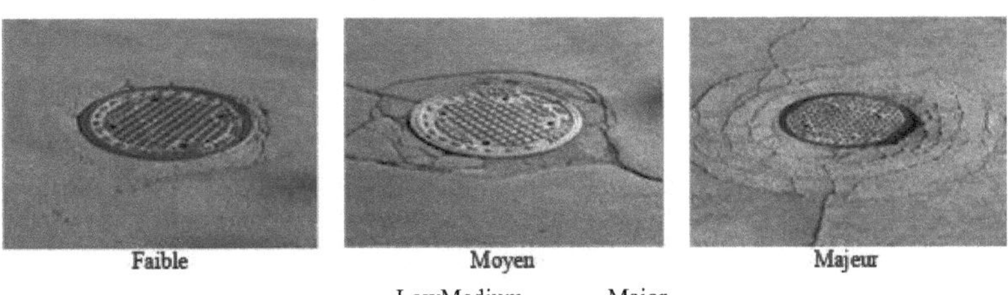

LowMedium   Major

Figure I.19 - *Cracking gravel around manholes and catch basins (MTQ-AIMQ 2002).*

Low: Simple, intermittent cracks with openings of less than 5 mm. The edges are generally clean and well defined;

Medium: Single cracks or multiple cracks along a main crack, the latter being open by 5 to 20 mm. The edges are sometimes eroded and slightly depressed. Without being uncomfortable, the crack is perceptible to the user;

Major: Single cracks or multiple cracks along a main crack, the latter being open by more than 20 mm. The edges are often eroded and there is subsidence or frost heave in the vicinity of the crack. Rolling comfort is reduced by surface deformation.

1.3.5.1.2   Probable causes

These cracks are produced by consolidation or settlement of the pavement, freeze-thaw cycles, desaggregation of the chimney by brine, dynamic impacts and loss of material around the structure.

1.3.5.2 Clearing out manholes and sumps

It's an unevenness between the surface of the lining and the top of a sump or manhole.

1.3.5.2.1 Size and scope

The three levels of severity of manhole and sump clearing are shown in Figure I.20. The extent is the total number of manholes per survey section and severity level.

Low: defined as a cleavage of less than 20 mm;

Medium: in this case, there is a cleavage of 20 to 40 mm;

Major: here the denivellation is more than 40 mm.

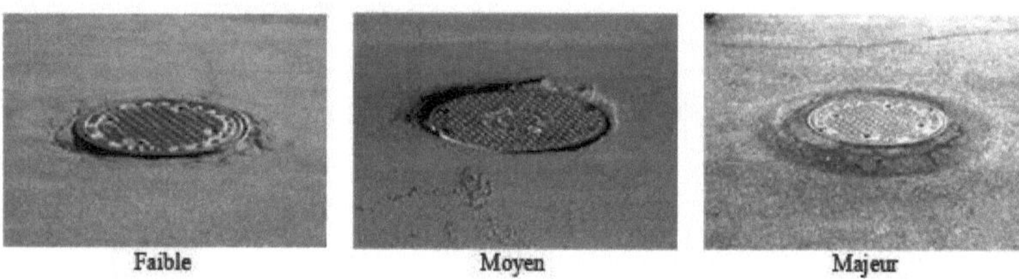

LowMedium Major

Figure I.20 *- Levelling gradient for manholes and catch basins (MTQ-AIMQ 2002).*

1.3.5.2.2 Probable causes

The most likely causes are as follows:

Consolidation or settlement of the carriageway;

Freeze and thaw cycles causing permanent deformation;

Disaggregation of the chimney in the presence of brine;

Dynamic impacts accumulating permanent deformations;

Loss of material around the structure.

1.3.5.3 Cutting and slicing

This is a crack or subsidence in the trench or in its vicinity.

1.3.5.3.1 Size and scope

The three levels of severity of manhole and sump clearing are shown in Figure I.21. The extent is the area affected over the section of the survey.

LowMedium Major

Figure I.21 *Cutting and slicing gravel (MTQ-AIMQ 2002).*

Low: this corresponds to a denivellation of less than 20 mm and/or a simple crack with an opening of less than 5 mm, the edges of which are generally clean and well defined;

Medium: this is a denivellation of 20 to 40 mm and/or single crack or multiple cracks along a main crack, the latter being open by 5 to 20 mm. The edges are sometimes rough and a little depressed;

Major: at this level the cleavage is more than 40 mm and/or single cracks or multiple cracks along a main crack, the latter being open by more than 20 mm. The edges are often eroded.

1.3.5.3.2 Probable causes

The most likely causes are as follows:

Insufficient compaction of the backfill materials in the trench;

The heterogeneity of trench materials and those of existing carriageways;

The release of stresses produced by a loss of lateral support in the slice;

Incomplete backfill under the edges of the paving ;

Lack of watertightness at the cutting joint.

I.4 Degradation of rigid pavements

When properly designed and laid, concrete slab pavements suffer little deterioration compared with bituminous pavements. The nature of this deterioration can take various forms, making it impossible to classify them into families. We will therefore deal with them individually, following the previously established schema. Thus, the most frequently encountered degradations on rigid pavements are: cracks, spalling, joint shifting, and pumping discharges or pumping phenomena.

I.4.1 Cracking

I.4.1.1 Description and development

There are two groups of cracks: *"longitudinal, transverse and oblique cracks"* and *"slab corner cracks"*. These are all slab fractures in two pieces, but the corner cracks correspond to an intersection of the crack with the edges of the slab, thus forming a triangle whose two right angles are the edges of the slab (Figure I.22).

The cracks are opening up and the levers are spalling, allowing material to escape and water to seep into the carriageway. As a result, the pounding of the slabs caused by heavy traffic also leads to pumping discharges. This opening is limited for continuous reinforced concrete until the reinforcement breaks. The crack may also branch out and the corners of the slabs adjacent to the crack may break.

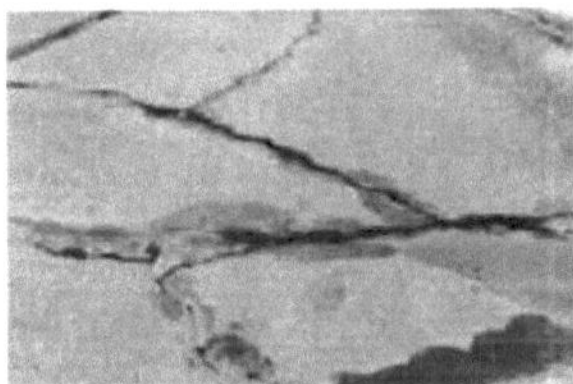

Cracks in various shapes Slab corner cracks
Figure I.22 - Crack network.

I.4.1.2 Probable causes

The probable causes of these cracks are :

Insufficient load-bearing capacity (slab too thin, insufficient tensile strength of concrete, etc.) ;

Degradation of slab support conditions (settlement or erosion of foundation soil) ;

Thermal shrinkage of concrete when sawn late;

Water shrinkage ;

Fatigue failure of the pavement due to the accumulation of excessive tensile and bending stresses;

Joint spacing too long ;

Swelling or shrinkage of the subsoil.

1.4.2   Spalling

1.4.2.1 Description and development

These are fragments that have detached from the concrete mass around joints or cracks. Generally, this degradation simply affects part of the thickness of the slab (Figure I.23). The spalls become more and more numerous and wider and evolve towards fragmentation into smaller and smaller slabs. They are also accompanied by slab collapse and material loss. Lastly, the phenomenon of water penetrating the carriageway is becoming more pronounced.

Figure I.23 - A spall on a rigid pavement.

1.4.2.2 Probable causes
The probable causes of this type of degradation are :
Blocked joints (presence of incompressible materials) preventing thermal expansion and creating compression at the edges, causing them to crumble;
Existence of areas of weakness at joints;
Insufficient compressive strength of concrete ;
Local deterioration of concrete by premature sawing;
Friction of joint lips caused by slab beating.

1.4.3   Joint offsets
1.4.3.1 Description and development
This is a vertical levelling between the two edges of a slab joint or the edge of a crack (Figure I.24). Joint misalignments lead to a change in the operating conditions of the slab (load transfer) and an alteration in the bond. They evolve towards spalling of the joint lips or crack edges, discharge of fines following water infiltration into the body of the pavement, as well as towards transverse or oblique cracking.

Figure I.24 - Slab offset of a rigid carriageway.

1.4.3.2 Probable causes
The probable causes of this type of degradation are :
Eroding foundation ;
Poor load transfer at transverse joints ;
Insufficient bearing capacity and/or cohesion of the supporting soil resulting in differential settlement;
Movement of materials under both edges of the joint due to pumping, water shrinkage or poor drainage.

1.4.4   Pumping
1.4.4.1 Description and development
This involves the ejection of materials (water, mud, etc.) onto the surface of the pavement when heavy vehicles pass over it, at cracks or joints due to the existence of cavities under the slabs (Figure I.25).
It evolves towards the formation of cavities in the ends of slabs. These cavities mean that heavy traffic causes the slabs to vibrate, accentuating pumping discharges. Eventually, this can lead to stair-

stepping and cracking of the slabs.

1.4.4.2 Probable causes

The probable causes of this type of degradation are :

Poor drainage of the carriageway;

Lack of cohesion and water sensitivity of the substrate;

Figure I.25 *- Material ejection (pumping) in a rigid carriageway.*

Degradation of slab bearing conditions in the presence of water due to dynamic stresses (slab movement under load generates pressurised water movements at the slab-foundation interfaces, causing water and fines to rise through joints or cracks).

I.5 Methods for assessing the behaviour of degraded pavements

However, in the case of deteriorated carriageways (after commissioning), a study must be undertaken to assess the state of deterioration and load-bearing capacity of carriageways, to survey the geometry of the road network and to ensure better knowledge of transport demand, its distribution and composition on the entire classified road network. To achieve this, summary inspections must be carried out on a regular basis in order to gather all the information needed to feed the database. Records are taken of the natural and economic environment of the road, as well as technical data on its state of deterioration, including previous maintenance work (reinforcement or periodic maintenance, in particular widening), average monthly rainfall over the last fifty years and the division into climatic zones.

According to the literature, there are several methods that can be used to characterise the condition of road damage. In this section, we will present three of the most commonly used methods:

The *"VIZIR"* method for surfaced roads;

The *"VIZIRET"* method for covered roads;

The *CEBTP-LCPC* method.

1.5.1   The "VIZIR" method

1.5.1.1 Principle of the "VIZIR" method

From the outset, the *"VIZIR"* method was a way of listing damage by reference to its extent and degree of severity. The recording of pavement deterioration, which was manual until 1988, is now computer-assisted thanks to the LCPC's *"DESY"* equipment. It is supplemented, in DESY, by software for calculating the surface index *"Is"*, which varies from *"1"* for the best pavements to *"7"* for the worst.

The VIZIR method goes much further than a simple list of deteriorations, especially as its ultimate aim is to assess the quality of road networks. The evaluation method is based on the following classic parameters of road testing: structure, maintenance, bearing capacity, evenness of the carriageway and local conditions. At the end of the chain, the VIZIR method is a scientific method for determining pavement maintenance and rehabilitation requirements.

1.5.1.2 Classification and quantification of damage

Through its classification and quantification of deterioration, the VIZIR method is designed to provide an image of the surface condition of a road at a given moment and to identify areas of equal quality classified into three levels of deterioration. These areas of equal quality and these three levels of deterioration are used to determine the nature and type of work required. The deteriorations listed in the VIZIR method mainly concern flexible bituminous pavements; they are classified into two categories: *"Type A"* and *"Type B"* deteriorations.

I.5.1.2.1 "Type A" structural degradation

They appear within the pavement structure or its support and affect the asset. They are degradations resulting from insufficient structural capacity of the pavement (Table I.1). They mainly include: (1) deformation and (2) fatigue cracking.

Table I.1 - Level of severity of type A deterioration (LCPC-VIZIR 1991).

Types of damage	Levels of seriousness		
	(1): Good condition	(2): Average condition	(3): Poor condition
Deformation and rutting	Sensitive to the user but not very important f < 2 cm	Severe deformation, localised subsidence or rutting 2 < f < 4 cm	Deformation seriously affecting safety or journey time f > 4 cm
Cracking	Fine cracks in the wheel tracks or axle	Clearly open and/or often branching cracks	Very branched and/or very open cracks; lips sometimes degraded
Faienfage	Fine trowelling with no material leaving a wide mesh (>50 cm)	Tighter soil (<50 cm) with occasional material loss, uprooting and potholes forming	Very open ditch, cut into slabs (<20 cm) with occasional material loss
Repair	Either repair of all or part of the carriageway or surface work related to type B defects	Surface work related to type A defects, satisfactory repair performance	Surface work related to type A defects. Damage to the repair itself

I.5.1.2.2 "Type B" surface damage

Also known as non-structural deterioration, it originates in the surface layer of the pavement and initially affects its superficial qualities. They give rise to repairs that are generally not related to the structural capacity of the pavement. Their origin is either a fault in laying, or a fault in the quality of a product, or a particular local condition which traffic can obviously accentuate (Table I.2).

Type B degradations include :

Cracking, excluding fatigue cracks, i.e. longitudinal joint cracks;

Transverse thermal shrinkage cracks, longitudinal or transverse clay shrinkage cracks (desiccation);

Pullouts ;

Movement of materials.

Table I.2 - Level of severity of type B deterioration (LCPC-VIZIR 1991).

Types of damage	Levels of seriousness		
	(1): Good condition	(2): Average condition	(3): Poor condition
Longitudinal joint cracks	Fine and unique	Wide (1 cm at most) without uprooting, finely branched	Wide with lip spurs or wide with branches
Potholes per 100m of carriageway	Quantity : < 5 Maximum size: Φ30	Quantity: 5 to 10 Size: Φ30 cm Or Quantity : < 5 Size: Φ100 cm	Quantity : > 10 Size: Φ30 cm Or Quantity : 5 to 10 Size: Φ100 cm
Types of damage	(1): Good condition	(2): Average condition	(3): Poor condition

Removal (stripping, plucking and peeling), movement of materials	Punctual without the appearance of the base coat	Continuous or punctual with the appearance of the base layer	Continuous with the appearance of the base layer
	Punctual	Continuous on one tread	Continuous on one tread and very marked

1.5.1.3 Survey and rating of damage

Damage is recorded by an operator who walks the route and makes notes:

Type of degradation ;

The severity of this deterioration;

The extent, i.e. the length of road affected or the surface area where applicable.

The VIZIR method provides a typology of damage and three levels of severity for each. However, the survey can be carried out *"manually"*, by walking along the road or in a car. The operator then reports his observations (identification of the damage and an estimate of its severity) on a route map, i.e. a document representing the line of the road, whose scale, and therefore precision, is appropriate to the type of study. The gravity values are average values that are suitable for many road networks; however, they can be modified according to the maintenance objectives set and the expected level of service. The VIZIR method therefore consists of considering several types of deterioration and repair work, assigning them a *"score"* according to the level of service encountered. The deterioration index is calculated mainly on a length of road from three groups of deteriorations, namely :

Group 1: cracks and faienfage ;

Group 2: Deformations and rutting ;

Group 3: repairs.

First of all, we calculate *a "cracking index: If"*, which depends on the severity and extent of the cracking or crazing along the length of road under consideration. Where there is both cracking and weakening, the higher of the two values is taken.

A *"deformation index: Id"* is then calculated, which likewise depends on the severity and extent of the deformation or rutting. The combination of the two indices *'If'* and *'Id'* gives an initial index that qualifies the pavement; if necessary, this can be corrected according to the severity and extent of certain repairs. In fact, some repairs mask a deficiency in the pavement and are therefore considered to be an aggravating factor when estimating surface quality.

After this correction, we end up with a *"surface index: Is"* which qualifies the pavement over the length chosen for the calculation. The overall rating of the surface index *"Is"* varies from *"1 to 7"*:

Grades *"1" and "2"* correspond to good surface conditions requiring no work (or only limited work), with few or no cracks or deformation;

Grades *"3" and "4"* correspond to fairly average surface conditions, bad enough to trigger maintenance operations apart from any other consideration; cracks with little or no deformation or deformation not accompanied by cracks;

Grades *"5", "6" and "7"* correspond to very poor surface conditions requiring major maintenance or reinforcement work; large quantities of cracks and deformation.

The method for determining the surface area index *"Is"* is summarised in the diagram below (Figure I.26). The base length on which the surface index *"Is"* is calculated may depend on the type of study, the database or other parameters used in the diagnosis, and the operator.

In the case of studies of road maintenance management support systems, which are global studies, the route map is drawn at a scale of approximately 2 cm per km; the database itself is constructed with a grid spacing of around 500 m. The *"Is"* can therefore be calculated with a 500 m grid.

In the case of a route maintenance project, the survey is carried out at a scale of approximately 5 graphic cm per km, so the *"Is"* can be calculated with a step of 200 m.

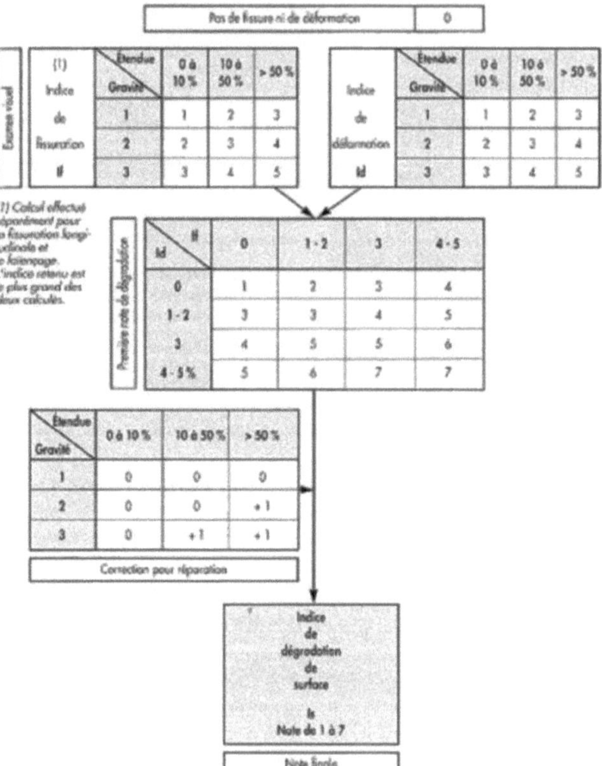

Figure I.26 - VIZIR method for assessing the condition of road pavements.

1.5.1.4 Searching for solutions

With regard to the search for solutions, the methodology of the studies conducted by the LCPC can be summarised as follows:

Building a database, collecting data in the field, determining the quality of the network at time t0;

Analysis of the data, determination of the "technical solution", i.e. what needs to be done to the road network in terms of its quality image over time to restore it to a certain level of service, without any budgetary constraints;

Search for an "optimised solution" that takes account of technical imperatives and budgetary constraints, including spreading the work over several years and projecting the image of the network at time t0 and the technical solution over time.

The technical solution is determined in two stages:

First, we associate *"visual condition"* and *"bearing capacity"*; bearing capacity is given by the value of the *"deflection: d"* and visual condition is represented by the surface degradation index *"Is"* determined earlier. From this combination, we derive a *"quality score: Qi"* for the pavement from *"1 to 9"* and a conclusion on what needs to be done: either nothing needs to be done, or maintenance or reinforcement needs to be carried out;

Then, in a second step, the *"quality rating"* of the carriageway is associated with the *"traffic level"*, and the solution for the works is determined for each pair of values.

1.5.1.4.1 Determination of the "Qi" pavement quality rating

The quality rating of the pavement is derived from the combination of the value of the *"Is"* index describing the surface of the pavement and that of the *"deflection" index* describing the overall bearing capacity of the pavement. In fact, the degradation index *"Is"* is classified into three categories as described above. However, deflection is classified into three categories determined by two thresholds *"d1"* and *"d2"*, namely :

The deflection *"d1"* is the value below which the structure is considered to behave satisfactorily;

The deflection *"d2"* is the value above which the structure is considered to have serious load-bearing defects;

Between *"d1" and "d2"* is the zone of indetermination.

The choice of *"d1"* and *"d2"* values depends on many factors, such as climate, type and thickness of road surfaces, soils and axle loads. These values are generally based on experience in a given country. The grid of pavement quality scores according to deflection and the "Is" index is presented in Table I.3. The quality scores mean the following:

- ₁, **Q2 and Q3**: these ratings mean that there is nothing to be done, or only maintenance work, the solution to which will be given later depending on traffic. When there is only sealing work to be considered, the cracking index is used to determine the date and nature of the work;

- ₇, **Q8 and Q9**: these notes mean that the carriageway requires reinforcement work, the thickness of which depends on the traffic. The solution is given later;

- ₄, **Q5 and Q6**: this is an area of indetermination that needs to be resolved by looking for the cause of the discrepancy between lift and visual examination. Schematically, this can be summarised as follows:

- **Q4**: pavement showing a pronounced state of deterioration despite good bearing capacity. The validity of the deflection should be checked (in particular the measurement period), as well as the nature of the deterioration (in particular the rutting of the asphalt layers which is not linked to the deflection). Depending on the response, Q4 will be reclassified as Q2 (priority to deflection) or Q7 (priority to the state of deterioration);

- **Q5**: same analysis as above; we'll take into account the position of the deflection in relation to the limits, as well as the traffic; depending on the answer, we'll reclassify as Q3, Q7 or Q8;

- **Q6**: pavement showing strong deflection with no apparent deterioration; to validate or invalidate the surface condition, we will check the age of the pavement or the date of the last works, as well as the level of traffic. Depending on the response, the road will be upgraded to Q3 or Q8.

Table I.3 - "Qi" quality rating (LCPC-VIZIR 1991).

Indice de dégradation Is \ Déflexion	d1	d2	
	Classe 1	Classe 2	Classe 3
1 – 2 Peu ou pas de fissures ou pas de déformations	Q_1	Q_3	Q_6
3 – 4 Fissuré sans ou avec peu de déformation et déformations sans fissures	Q_2	Q_5	Q_8
5 – 6 – 7 Fissures et déformations	Q_4	Q_7	Q_9

1.5.1.4.2 Determination of the solution

Determining the solution goes beyond the strict scope of the VIZIR method, which is a method for determining the quality of a road network and the work required to restore it, and not a method for calculating pavement reinforcement. The VIZIR method is only intended to situate the place of the *"Is"* deterioration index and the *"Qi"* quality rating of the pavement in the process of choosing a solution.

However, the solution is determined by crossing the traffic class and the pavement quality rating. As an indication, we give the example of a solution that was established for the study of the implementation of a road maintenance management support system in Korea (Table I.4).

Table I.4- Example of the solution table as a function of traffic (LCPC-VIZIR 1991).

Traffic Quality rating		0 a 500 T4	500 a 1000 T3	1000 a 2000 T2	>2000 T1
Maintenance	Qi	TS	TS	4BB	5BB
	Q2	TS	TS or 4BB	4BB	5BB
	Q3	TS or 4BB	4BB	7BB	7BB
Renforceinent	Q6	7BB	7BB	10BB	7ED-5BB
	Q^7	7BB	10BB	7ED+5BB	-
	Q8	10BB	7ED+5BB	10ED-5BB	15ED+5BB

1.5.2 The "VIZIRET" method

The "VIZIRET" method is based on a process which is entirely identical to that of VIZIR, except that it takes account of other types of degradation, namely :

Corrugated sheet ;

Deformations ;

Ravine ;

Potholes

Based *on* the indices of the various degradations, we give the road the SQI (Structural Quality Index) viability index, which is equal to the maximum degradation index of the four families mentioned above.

1.5.3 The *CEBTP-LCPC* method

The other French method of testing carriageways is the *CEBTP-LCPC* method, which applies to both network and road studies. Network studies consist of a statistical evaluation of the quality of the structures and a determination of the maintenance and repair work required. By assessing the quality of carriageways, they enable a maintenance, repair and route improvement strategy to be defined, depending on traffic conditions. The route studies assess the characteristics of the structures and solutions applicable to homogeneous trunks. Restoration solutions range from one-off repairs to continuous reinforcement over long stretches of carriageway, including improvements to geometry and comfort.

I.5.3.1 Methodology

The investigations in both cases (network and route studies) are similar and are based on the following points:

The history of the chaussee ;

Damage survey;

Deflection measurements.

These three points are treated under the same conditions as the VIZIR method. Three categories of carriageway sections are identified on the basis of the deterioration surveys:

Sections in apparent good condition;

Cracked or meshed sections ;

More or less deformed sections.

To quantify the apparent quality of pavements, a notation is adopted which compares the percentage of the route line affected by deterioration to the length of the unit section taken into account (for example 500m or 1000m). This notation is as follows (Table I. 5):

The mark of "1" is for less than 10% of degradation;

The mark of "2" is for 10% to 50% degradation;

A score of "3" is more than 50% degradation.

Table I.5 - Grading based on the degree of deterioration of the pavement (CEBTP-LCPC 1985).

Fissures / Déform.	1	2	3
1	1	2	3
2	3*	4	5
3	5*	6	7

rarest cases 40

By combining the two types of deterioration (cracks and deformations), we obtain the grid below, which gives a figure for the apparent quality of a section of pavement (Figure I.27).

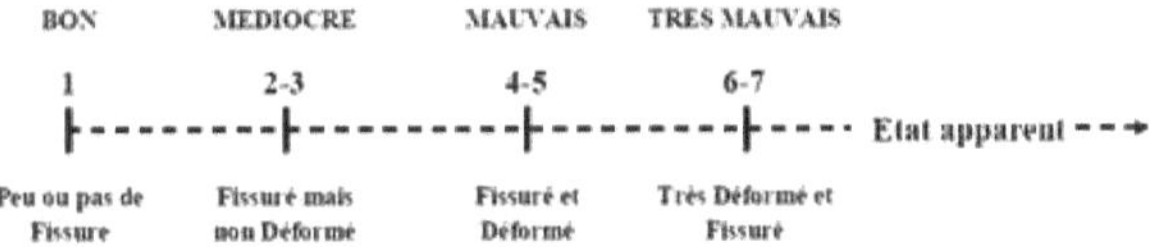

Figure I.27 - Apparent quality of a pavement section (*CEBTP-LCPC 1985*).

This method disregards defects specific to the wearing course, such as ravelling, combing, bleeding, creep, etc., which are analysed separately and are the subject of appropriate remedial solutions.

I.5.3.2 Calculation of results

The calibration of the results is a process which aims, on the basis of the characteristic deflection thresholds specific to the route or network, to compare the apparent state and deflection parameters with a view to producing a decision and solution selection grid as shown below (Table I. 6):

Table I.6 - *"Qi" quality rating* (*CEBTP-LCPC 1985*).

Etat apparent	Déflexion	Faible d1		d2 Elevée
Bon	1	Q1	Q2	Q3
Fissuré non déformé sans	2-3	Q2	Q3	Q4
Déformé et fissuré	4-7	Q3	Q4	Q5

I.6 Conclusion

This chapter has made it possible to present the vast majority of degradations on surfaced pavements. Thus, for bituminous surfacing, we have been able to study all types of deterioration in the form of families. In addition, we have discussed the main methods for assessing the deterioration of pavements themselves. In the next chapter, we will present the various road tests carried out on soils (construction materials for different road pavement layers).

CHAPTER II

ROAD IDENTIFICATION TESTS

11.1 Introduction

After presenting, in Chapter I, the main types of deterioration observed on bituminous surfacing and their probable causes, we will now briefly present, in this chapter, the main identification tests on materials used in road pavement structures.

11.2 Identification tests on materials used in road pavements

On the one hand, the mechanical behaviour of a pavement (before it is put into service) can be evaluated and checked to ensure its performance, safety and durability throughout its life. On the other hand, the quality of the materials making up the road pavement structure is also a fundamental parameter that requires prior study before these materials are used.

In fact, the tests which are concerned with the correct choice of materials suitable for the rules of embankment construction (classification of materials) and thus with the correct mechanical behaviour of the road pavement structure can be classified into three groups presented below. It should be noted that for each of these tests we will present only the objective of the test, the principle, the material used, the parameters estimated and/or calculated and the expression and interpretation of the results.

a) Identification tests

Stain test or "Methylene blue";

Particle size analysis ;

The limits of Atterberg ;

Cleanliness test or "sand equivalent" ;

Compaction test or "normal or modified Proctor".

b) Hardness tests

Los Angeles test drive ;

Micro-Deval test ;

c) Load-bearing capacity tests

Plate test;

C.B.R. (California Bearing Ratio) test.

II.2.1 Identification tests

II.2.1.1 Methylene blue stain test

11.2.1.1.1 Aim of the test

The clay minerals present in soils are mainly derived from the physicochemical alteration of rocks. The layered crystalline structure of clays gives them a set of behavioural properties linked to their affinity for water, *known as "activity"*, which leads to the swelling, plasticity and cohesion phenomena observed in these soils.

11.2.1.1.2 Principle of the test

The methylene blue test is used to assess the overall activity of the clay fraction of a soil by measuring the internal and external surfaces of the clay grains. To do this, molecules of methylene blue are fixed to the clay grains, and a simple test is used to assess the amount of blue fixed. This is used to calculate the soil blue value or VBS, which is an essential indicator for classifying the soils used for earthworks and road bases.

11.2.1.1.3 Equipment used

Balance of sufficient range with a relative accuracy of 0.1%;

Stopwatch or equivalent indicating the second;

Square mesh sieve with 5 mm opening;

Plastic or graduated glass beaker;

A mechanical paddle stirrer with a speed of rotation covering at least the range 400 rpm to 700 rpm. The diameter of the blades is between 70 mm and 80 mm. The shape and dimensions of the blades must allow all the soil particles to be set in motion;

A cylindrical container (glass, plastic, stainless steel) with a minimum capacity of 3000 cm3 and an internal diameter of (155 ± 10) mm;

50 ml burette, or an automatic burette, graduated in 1/10th ml ;

White filter paper with a weight per unit area of (95 ± 5) g/m^2 , a thickness of (0.2 ± 0.02) mm, a filtration speed of (75 ± 10) s per 100 ml (according to the ASTM method) and a retention diameter of (8 ± 5) ^m ;

Glass rod 8 mm in diameter and approximately 300 mm long;

Solution of medicinal grade methylene blue at 10 g/l plus or minus 0.01 g/l (shelf life: 1 month maximum);

Demineralised or distilled water.

11.2.1.1.4 Operating mode

Dosing consists of successively injecting specific doses of methylene blue into the soil suspension until saturation of the clay particles is reached. The spot test is used to identify the moment of saturation. A drop of liquid is taken from the beaker containing the blue-soaked soil, and placed on the filter paper held horizontally in the air (diameter of the deposit between 8 and 12 mm). There are two possible cases:

If the central spot is surrounded by a turquoise blue aureole, the test is positive. In this case, the test is complete and the clay particles are saturated with Methylene Blue. The test is repeated identically, five times at one-minute intervals to confirm;

If the stain is surrounded by a colourless moist aureole, the test is negative. In this case, methylene blue is added in 5 cm3 doses until the test is positive. The test is repeated identically, five times at one-minute intervals to confirm.

Table II.1 summarises the detailed phase-by-phase procedure for carrying out the methylene blue test.

Table II.1 - Procedure for carrying out the methylene blue test.

Cinematic dosing	Comments
a. Add 5 cm3 of blue then go to 2.	1ere phase: addition of methylene blue in coarse steps (5 cm3) followed by the stain test after 1 min ± 10 s.
b. Spot test after 1 min t 10 s : - If the test is negative, return to 1; - If the test is positive, go to 5.	If the test is positive with less than 10 cm3 of methylene blue, repeat the test with a larger sample.
c. Add 2 cm3 of blue then go to 4.	2eme phase: after the first immediate positive test, the blue is added in fine steps (2 cm3) because the suspension is becoming saturated.
d. Immediate spot test after 1 min: - If the test is negative, go back to 3 ; - If the test is positive, go to 5.	/
e. Perform the test confirmation 5 times every minute for 5 minutes: - If the test is negative, return to step 3; - If the test is positive, end of dosage.	3eme phase: confirmation of positive test for 5 minutes.

11.2.1.1.5 Expressing results

Water content of the sample :

$$w = \frac{m_{h2} - m_{g2}}{m_{g2}} \qquad (\%) \qquad (\text{Eq. II.1})$$

Dry mass of the test sample :

$$m_0 = \frac{m_{h1}}{1 - w} \qquad (\%) \qquad (\text{Eq. II.2})$$

Mass of blue introduced (10 g/l solution) :

$$B = 0.01 * V \qquad \text{(\%)} \qquad \text{(Eq. II.3)}$$

Blue value for materials with Dmax < 5 mm :

$$VBS = \frac{B}{m_0} * 100 \qquad \text{(\%)} \qquad \text{(Eq. II.4)}$$

Blue value for materials with Dmax > 5 mm :

$$VBS = \frac{B}{m_0} * C * 100 \qquad \text{(\%)} \qquad \text{(Eq. II.5)}$$

Or :

VBS: is the methylene blue value of a soil. It is expressed in grams of blue per 100 g of the 0/50 mm fraction of the soil studied;

Dmax: is the maximum dimension of the largest elements contained in the soil (see NF P 11-301);

mh1: is the wet mass of the sample constituting the first test sample (expressed in grams) ;

mh2: is the wet mass of the sample intended to be dried, constituting the second test sample (expressed in grams) ;

mh3: is the wet mass of the sample making up the third test sample (expressed in grams) ;

ms2: is the mass of the sample after drying, corresponding to the second test sample (expressed in grams);

B: is the mass of blue introduced into the solution (10 g/l solution);

V: is the volume of the blue solution used (expressed in cubic centimetres);

C: is the proportion of the 0/5 mm fraction in the 0/50 mm fraction of the dry material;

w: is the water content, expressed as a decimal value.

11.2.1.1.6 Interpretation of results

Table II.2 shows the soil type classification according to the soil blue value coefficient (VBS) (LCPC-SETRA 2000).

Table II.2 - *Classification according to VBS (SETRRA-LCPC 2000).*

VBS (%)	Classification
0.1	Soil insensitive to water
0.2	Onset of water sensitivity
1.5	Threshold distinguishing sandy-loam soils from sandy-clay soils
2.5	Threshold for distinguishing low plasticity silty soils from medium plasticity silty soils;
6	Threshold distinguishing silty soils from clayey soils
8	threshold for distinguishing clay soils from very clay soils

II.2.1.2 Particle size analysis

Granulometric analysis enables aggregates to be distinguished according to the granular classes marketed by manufacturers. The development of a concrete composition requires a perfect knowledge of the granulometry and granularity, as the strength and workability of the concrete depend essentially on the aggregate. Furthermore, the *"D"* dimension of the aggregate is limited by various considerations concerning the structure to be concreted: thickness of the part, spacing of the reinforcement, density of the reinforcement, complexity of the formwork, risk of segregation, etc.

The sands must be of a grain size such that the fine elements are neither in excess nor in too small a proportion. If there are too many fine grains, it will be necessary to increase the water content of the concrete, whereas if the sand is too coarse, the plasticity of the mix will be insufficient and will make placing difficult.

Aggregates used in building and civil engineering are rolled or crushed materials of natural, artificial or recycled origin, with dimensions between 0 and 125 mm. They are not generally made up of elements of equal size, but of a set of grains whose varying sizes fall between two limits: the smallest "*d*" and the largest "*D*" dimension in mm.

Aggregates are sorted by size using *"square mesh"* sieves and *"circular hole"* sieves and a class of aggregate is designated by one or two figures. If only one figure is given, it is the maximum diameter *"D"* expressed in mm; if two figures are given, the first indicates the minimum diameter *"d"* of the grains and the second the maximum diameter *"D"*. There are five main granular classes characterised by the extreme dimensions *(d and D)* of the aggregates encountered:

0/D "fines" with D <0.08 mm ;

0/D "sands" with D <6.3 mm ;

Gravel" d/D with d >2 mm and D <31.5 mm ;

Pebbles" d/D with d >20 mm and D <80 mm ;

The "low" d/D with d >6.3 mm and D <80 mm.

11.2.1.2.1 Aim of the test

Granulometric analysis by *"sieving"* and *"sedimentometry"* consists of determining the size distribution of the grains making up a granulate with dimensions between *"0.063"* and *"125 mm"* (determination of the distribution of soil grains according to their size in a sample). It is called :

- A *"Refusal"* on a sieve is the quantity of material retained on the sieve;
- A *"Sieve"* is the quantity of material that passes through the sieve.

Representation of the distribution of the mass of particles in the dry state as a function of their size.

11.2.1.2.2 Principle of the test

Granulometric analysis by *"sieving"* consists of dividing a material into several granular classes of decreasing size using a series of *"sieves"*. The masses of the different rejects and sieves are related to the initial mass of the material. The percentages thus obtained are analysed graphically.

However, *"sedimentometry"* is a test that completes the granulometric analysis of soil by sieving. It applies to particles with a diameter of less than 0.100 mm. This analysis is used to determine the percentage of clay particles in a material. In the study of soils, clay is defined as the fraction of material comprising elements with diameters smaller than 2um, and separated during the mechanical analysis.

Granulometric analysis using *"sedimentometry"* involves measuring the sedimentation time in a water column, i.e. the speed at which particles fall.

11.2.1.2.3 Equipment used

Screening machine (Vibro-tamis) ;

Lid to prevent loss of material during sieving and a bottom receptacle to collect the last sieve;

Metal or plastic containers;

Hand scoop for filling ;

5 kg capacity, 1 g accuracy ;

Oven set at 105 ± 5°C;

Series of sieves with standardised mesh openings ;

The mesh size and number of sieves are chosen according to the nature of the sample and the expected precision;

The recommended sieve openings for measuring the size of sand aggregates are (in mm) :

- The series of sieves 0.08 - 0.16 - 0.32 - 0.63 - 1.25 - 2.5 - 5 - 10 - 20 - 50 - 100 - 200 is adopted by the former French standard XP P18-540 (1997);
- The current standard (EN 933-2 2020) recommends the following series of sieves for the particle size analysis of sands: 0.063 - 0.125 - 0.25 - 0.50 - 1 - 2 - 4 - 8 - 16 - 31.5 - 63 - 125.

The recommended sieve openings for measuring the size of gravel and pebbles are (in mm) :

- The 8 - 16 - 31.5 - 63 - 125 sieve series can be used in accordance with standard EN 933-2 (2020);

- The 6.3 - 8 - 10 - 12.5 - 16 - 20 - 25 - 31.5 - 40 - 50 - 63 - 80 sieve series is also available.

11.2.1.2.4 Operating mode

Firstly, the sample analysed must be large enough to be measurable and not too large to saturate the sieves or cause them to overflow. It is therefore unthinkable to analyse a sample weighing just one microgram, in the same way as a sample weighing one tonne. The *'M'* mass limit range avoids these problems. The mass of the sample to be taken, *"M"*, must be within the range: *"0.2D < M < 0.6D"*. Note that this range is expressed as a function of *"D"*, which represents the *"D"* of the aggregate class *"d/D"* in mm or *"M"* is indicated in kg. For example, to carry out a granulometric analysis of a 4/12.5 gravel, you need to identify *"D"*: in this case *D=12.5*mm, you will then need to take samples with a mass between: 0.2*12.5 < M < 0.6*12.5, i.e.: 2.5kg < M < 7.5kg. Here, the mass *"M"* can be chosen to be equal to 3kg. The following steps should then be followed for particle size analysis by sieving:

Oven dry at 105 ± 5°C for 24 hours;

Take a quantity of dry material, which depends on the maximum grain size "D";

Weigh the mass "D" of material within the limits defined by the following formula: 0.2D < M < 0.6D where mass "M" is expressed in (kg);

Assemble and fit the column of sieves in descending order of mesh opening, then add the lid and the watertight base to collect fine particles;

Pour the granules onto the upper sieve and place the lid on;

Attach the series of sieves to the vibrating screen and vibrate them for a few minutes. Finish by shaking horizontally and vertically;

Take the top sieve alone with its contents and shake it on a clean tray. Stop shaking when the sieve reject does not vary by more than 1% by mass per minute of sieving;

Weigh the reject (to within 0.1 %) and pour the sieve onto the next sieve with what is already there;

Do the same with the second sieve. Place the new reject on the balance with the first and pour the new sieve onto the third sieve. Note the cumulative rejection of the two sieves;

Sieve through to the last sieve. By weighing the sieved material in the bottom with the sum of the accumulated rejects, find the mass weighed at the start. The loss of material must not exceed 2% of the total mass of the sample;

Plot the particle size distribution curve on a graph with the percentage of sieved material under the sieves whose mesh size "D" is indicated on the x-axis using a logarithmic scale. For example, to trace the particle size distribution curve for 0/5 sand, the weight of the successive sieves is used to determine the percentages of the sieves (see table below) corresponding to each of the sieves used.

For granulometric analysis by sedimentation, a 20g mass of dry material is weighed and placed in a test tube *"A"*. Next, 30cm^3 of 5% sodium hexametaphosphate solution and 200cm^3 of demineralised water were added, and the whole mixture was shaken manually. After resting for 24 hours, the solution is stirred again for 10 minutes and the volume of the solution in test tube *"A"* is made up to 1000cm^3 with demineralised water.

In the same way, another solution is prepared in test tube *"B"*: 30cm^3 of 5% sodium hexametaphosphate solution, topped up to 1000cm^3 with demineralised water. Test tubes *"A"* and *"B"* are placed in a cold water bath to even out the temperature. The densimeter is first introduced into test tube *"B"*, and at the same time the solution in test tube *"A"* is shaken vigorously.

At different times, after taking the reading from *"B"*, the densitometer is removed and gently inserted into the *"A"* test tube and the reading taken again. Each time the densimeter is read, the temperature is also measured in the cold water bath.

11.2.1.2.5 Expressing results

The **uniformity coefficient** *"Cu"* is used to describe the granulometry which is calculated by the equation below as follows:

$$C_u = \frac{D_{60}}{D_{10}} \qquad \text{(sans unité)} \qquad \text{(Eq. II.6)}$$

The **coefficient of curvature** *"Cc"* is calculated by the equation :

$$C_c = \frac{D_{30}^2}{D_{10} * D_{60}} \qquad \text{(sans unité)} \qquad \text{(Eq. II.7)}$$

The **fineness modulus** *"Mf"* is an important characteristic, especially for sands. By definition, it is the hundredth (1/100) of the sum of the cumulative refusals (expressed as a percentage by mass) of the sieves in the following series: 0.125 - 0.25 - 0.5 - 1 - 2 - 4.

$$M_f = \frac{\sum Refus\ cumulé\ (0.125 + 0.25 + 0.5 + 1 + 2 + 4)}{100} \qquad (\%) \qquad \text{(Eq. II.8)}$$

11.2.1.2.6   Interpretation of results

The result of the **uniformity coefficient** "Cu" can be interpreted as follows (Table 11.3)   :

Table II.3 - Granulometry as a function of Cu.

Cu	Granulometry
1	A single lump
1 a 2	Very uniform
2 a 5	Uniform
5 a 20	Little uniformity
> 20	Very well

The result of the **curvature coefficient** "Cc" can be interpreted as follows: It is considered that when Cu is greater than 4 for gravel and greater than 6 for sand, then 1<Cc<3 gives a well graded particle size (low porosity) or well graded material (continuity is well distributed). If Cc < 1 or Cc > 3 -* poorly graded material (continuity is poorly distributed).

The result of the **finesse module** can be interpreted as follows:

•   For 1.8 < Mf < 2.2: sand contains a majority of fine and very fine elements, which means that the water content needs to be increased. sand should be used if ease of processing is particularly important, to the probable detriment of strength;

•   For 2.2 < Mf < 2.8: sand should be used if satisfactory workability and good resistance are required with limited risk of segregation. It is a good sand;

•   For 2.8 < Mf < 3.2: the sand lacks fineness and the concrete loses workability. Sand should be used if high strength is required at the expense of workability and with the risk of segregation;

•   For Mf > 3.2 the sand must be rejected.

<u>*Note*</u>: The correction of an aggregate is necessary when its particle size curve shows a discontinuity or when there is a lack or an excess of grains in a sieve zone. Correction consists of compensating for these deviations by adding another aggregate until a mixture with the desired qualities is obtained. This practice is usual for modifying the fineness modulus *"Mf"* of hydraulic concrete sands.

11.2.1.3 Atterberg limits

Atterberg limits are water contents corresponding to particular states of a soil. This test is generally applied to soils with a percentage of fines (80цш) greater than 35%. On the other hand, the determination of the clay content of a soil using the Atterberg limits rather than the VBS test (Value of Soil Blue, VBS) is preferable whenever the soil is clayey to very clayey.

11.2.1.3.1   Aim of the test

The main objective is to characterise the clay content of a soil, and therefore to determine the remarkable water contents located at the border between these different states are the "Atterberg Limits", i.e. :

The liquid limit (wL), which defines the boundary between the plastic and liquid states; The plastic limit (wP), which defines the boundary between the solid and plastic states.

11.2.1.3.2 Principle of the test

The test is carried out on the 0/400µm fraction in two phases:

Determination of the water content "WL" at which a groove in a cup closes to 10 mm after 25 repeated impacts (this liquid limit corresponds to conventional shear strength);

Determination of the "WP" water content at which a 3 mm diameter roll of soil will crack (this plasticity limit corresponds to conventional tensile strength)

11.2.1.3.3 Equipment used

Casagrande device;

Grooving tool and spatula ;

10 mm thickness shim ;

Marble top with hairdryer;

Weighing capsules ;

Oven ;

Balance ;

Pissette.

11.2.1.3.4 Operating mode

The material prepared for the test must be kneaded to obtain a homogeneous, almost fluid paste. In order to determine the liquid limit, a clean, dry dish is filled with a spatula and a mass of paste of approximately 70g. This paste, spread out in several layers to avoid trapping air bubbles, has a symmetrical appearance at the end of the operation with respect to the vertical axis of the cup. The dough is then divided into two parts using the grooving tool. The cam is activated to subject the cup to a series of impacts at a rate of 2 blows per second. The number of impacts (NC) required to bring the groove lips together over a length of approximately 10 mm according to NF P 94-051 (1993) or over a length of approximately 13 mm according to ASTM D4318 (2000) is recorded. The entire operation is carried out at least four times on the same dough, but with a different water content each time. The test is only continued when NC is between 15 and 35. The number of shocks in the series of tests must be within 25 and the difference between two consecutive values must be less than or equal to 10. Finally, approximately 5 to 10 g of dough is removed from the cup using a spatula on each side of the groove lips to determine the water content by oven drying.

By definition, the liquid limit *"WL"* is the water content of the material which conventionally corresponds to a closure of 10 mm or 13 mm (depending on the AFNO or ASTM standard used) of the lips of the groove after 25 impacts. It is calculated from the equation of the mean straight line fitted to the pairs of experimental values (NC, WL) for at least four pairs of values. The WL is obtained for a NC value equal to 25 blows. It is expressed as a percentage and rounded to the nearest whole number.

However, for the plasticity limit, a pellet of the previously prepared dough is formed and rolled on a smooth plate by hand so as to obtain a roll that is gradually thinned until it reaches 3 mm in diameter. The plastic limit is reached when, at the same time, the roll cracks and reaches a diameter of 3 mm. Once the cracks have appeared, the central part of the roll is removed and placed in a capsule of known mass, weighed immediately and placed in the oven to determine its water content.

By definition, the plasticity limit *"WP"* is the conventional water content of a roll of soil that cracks when its diameter reaches 3 mm. This plasticity limit is the arithmetic mean of the water contents obtained from three tests. The value is expressed as a percentage.

11.2.1.3.5 Expressing results

According to Atterberg (1911), the plasticity index *"PI"* is the range of water contents within which the soil behaves like a plastic material. The plasticity index is therefore equal to the difference between the values of the liquid limit and the plastic limit.

$$I_p = W_L - W_P \qquad (\%) \qquad (\text{Eq. II.9})$$

The Atterberg limits are used to calculate the consistency index "Ic", which characterises the water status of a soil (W_n: is the natural water content):

$$I_c = \frac{W_L - W_n}{I_p} \qquad \text{(sans unité)} \qquad \text{(Eq. II.10)}$$

Ic = 0 if Wn = WL: the material is in a liquid state;
Ic = 1 if Wn = WP: the material is solid.

11.2.1.3.6   Interpretation of results

The GTR uses the following thresholds (Table II.4):

Table II.4 - Argillosite as a function of the plasticity index (LCPC-GTR 2000).

Ip	Argilosite
0 a 12	Low
12 a 25	Average
25 a 40	Strong
> 40	Very strong

Also, the plasticity index characterises the width of the zone where the soil under study has plastic behaviour (Table II.5):

Table II.5 - State of the iol as a function of the plaiticity index (LCPC-GTR 2000).

Ip	Soil condition
0 a 5	Non plastic
5 a 15	Little plastic
15 a 40	Plastic
> 40	Very plastic

11.2.1.4 Cleanliness test or "sand equivalent

The sands used in different fields are not all clean; they contain a greater or lesser proportion of harmful fine clays which can considerably reduce the quality of the materials. This relative proportion of impurities in the sand can be determined using a cleanliness test known as the "sand equivalent". This test consists of flocculating the impurities in the sand under standard conditions of time and agitation (EN 933-8-1999; NF EN 933-8-1999).

11.2.1.4.1   Aim of the test

The test is generally used to measure the cleanliness of the sand used in the composition of concrete. The standardised procedure makes it possible to determine a sand equivalent coefficient which quantifies the cleanliness of the sand.

11.2.1.4.2   Principle of the test

The test consists of suspending the fine *"particles <0.063 mm or 63 pm"* after agitation, then leaving them to settle at the bottom of a transparent tube as shown in Figure II.1.

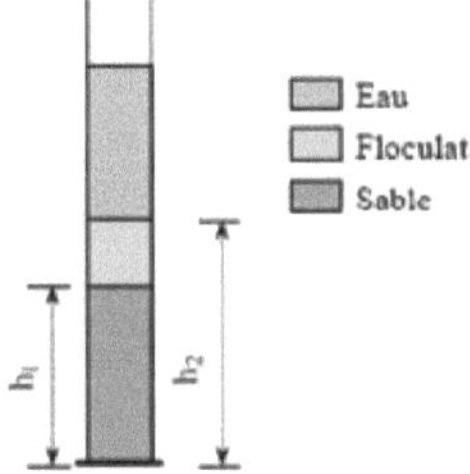

Figure II. 1 - Sand equivalent test (LCPC-GTR 2000).

11.2.1.4.3   Equipment used

The main items of equipment are two graduated test tubes, in which the test will be repeated in a similar way, a piston with a well-defined weight and an automatic stirrer. The test tubes are made of glass or transparent plastic, 40 cm high, fitted with a rubber stopper and graduated. The measuring piston consists of :

A 43 cm long rod;

A base with a diameter of 2.5 cm, the lower surface of which is flat, smooth and perpendicular to the axis of the rod and which has three screws at the side for centring the piston in the cylinder;

A sleeve, 1 cm thick, which fits over the graduated cylinder and guides the rod, at the same time as being used to locate the insertion of the test piston into the cylinder;

A weight fixed to the upper end of the rod to give the entire test piston, excluding the sleeve, a total mass of 1 *kg* ;

A wash tube is also used, with a length of 50 cm and an internal diameter of 4 *mm*. It is used to circulate the washing solution through the sample to be tested.

Plastic test tubes (glass) standardised "with two (2) reference marks" fitted with two stoppers;

Normalized tare piston;

Washing tube, Funnel ;

5-litre carboy with piston and 1.50 m flexible tube;

Non-specialised equipment for everyday use can also be used, such as: a measuring ruler, a sieve, a spatula, various containers, a balance and a stopwatch.

11.2.1.4.4 Operating mode

The test is carried out on the 0/2 mm fraction of the sand to be studied. The sample is washed using a standard process and left to stand. After 20 minutes, the elements are measured. This can be done using two types of measurement:

Measurement with a scale (ESV) ;

Measurement with a piston (ES).

a)- Preparation of test equipment

Sieve the sand with a 4 mm sieve, remove the reject, and collect all the sieve;

Place the carboy containing the 5 litres of washing solution a (1 *m)* above the bottom of the test tubes;

Prime the sound absorber and connect it to the scrubber tube;

Prepare two clean standard test tubes;

Fill both test tubes with washing solution up to the 1st mark (lower line).

b)- Filling, stirring and washing

For the three tests, fill the graduated cylinders with a washing solution up to the first lower mark;

Using a funnel, pour 120 *g* of dry sand into the test tube. If no dry sand is available (oven-dried at 105 ± 5°C for 24 hours), determine the water content (w) and take a quantity of wet sand corresponding to 120 g of dry sand, i.e.: 120 (1 + w) *(in g)* ;

Eliminate air bubbles (tap against the palm of the hand);

Leave for 10 minutes to moisten the test tube;

Stopper the test tubes, and shake each test tube for 30 seconds using an electric shaker (rectilinear, horizontal, sinusoidal movement, 20 cm amplitude, 90 back and forth strokes);

Wash and fill the test tubes in an upright position, using the wash tube:

• Rinse the stopper over the test tube and lower the washing tube into the test tube so that it passes through the sediment at the bottom of the cylinder, rotating it between the fingers, thus washing the inside walls of the test tube;

• To wash the sand, slowly lower and raise the washing tube which is turned between the fingers in the mass of sand, bringing up the fine particles and clayey elements;

• Pull out the wash tube slowly when the liquid level reaches the upper line.

Leave each test tube to stand for 20 minutes, avoiding any vibration.

11.2.1.4.5 Expressing results

The proportion of fines in relation to the rest of the sample is then measured and calculated by Eq. II.11 below:

$$E_S = \frac{H_1}{H_2} * 100 \qquad (\%) \qquad (Eq.\ II.11)$$

Or :

H_1: height of clean sand + fine elements (or flocculants) ;

H_2: total height of clean sand only.

11.2.1.4.6   Interpretation of results

The calculation is made for each test piece. If the two values obtained differ by more than four "4", the test procedure must be repeated. The sand equivalent (SE) of the sample tested is the average of the values obtained for each test piece, rounded to the nearest whole number. The cleanliness recommendations for sands used in concrete are given in Table II.6.

For road concrete, whether swept, streaked, printed, desactivated or corked, the recommendations are as follows: (ES>60).

For fine sand: if 60< ES <70, the sand is said to be clayey.

For coarse sand: if ES >80, the sand is said to be very clean;

Table II.6 - Nature and recommended values for sand equivalent.

ES at piston (%)	Nature and quality of the sand
ES<60	Clayey sand - Risk of shrinkage or swelling, to be rejected for quality concrete.
60 <ES<70	Slightly clayey sand - suitable for quality concrete where there is no particular fear of shrinkage.
70 < ES < 80	Clean sand - low percentage of clay fines Perfect for high quality concrete.
ES >80	Very clean sand - the almost total absence of fine clay may lead to a lack of plasticity in the concrete, which will have to be compensated for by increasing the water content.

11.2.1.5 Compaction test

The aim of soil compaction is to improve the geotechnical properties of the soil. It depends on four main variables:

The density of dry soil ;

Water content ;

Compaction energy ;

Soil type (particle size distribution, presence of clay minerals, etc.)

11.2.1.5.1   Aim of the test

The purpose of the Proctor test is to determine the optimum water content for a given backfill soil and fixed compaction conditions, which leads to the best possible compaction or maximum load-bearing capacity.

11.2.1.5.2   Principle of the test

The test consists of compacting the soil sample to be studied in a standard mould, using a standard tamper, according to a well-defined process, and measuring its water content and specific dry weight after compaction. The test is repeated several times on samples with different water contents. Several points on a curve (w; Yd) are thus defined; this curve is plotted as a maximum, the abscissa of which is the optimum water content and the ordinate the optimum dry density. Two types of moulds can be used for these tests, depending on the fineness of the soil grains:

The Proctor mould $\Phi_{moule\text{-}mt\grave{e}rieu\text{-}}$ = 101.6 mm / H = 117 mm (without extension), $V_{moule\text{-}Proctor}$ = 948cm^3 ;

The CBR mould Φ_{mould} = 152 mm / H = 152 mm (without riser) with a 25.4 mm thick spacer disc, i.e. a $_{Hutile}$ height = 126.6 mm, $V_{mould\text{-}CBR}$ = 2296 cm^3 .

Two types of test can be carried out with each of these moulds (depending on the compaction energy required):

The Proctor Normal (PN) test;

The Modified Proctor Test (MPT).

The choice of compaction intensity depends on the overload that the structure will be subjected to during its lifetime:

Normal Proctor test: Relatively low desired strength, of the unloaded or lightly loaded backfill type;

Modified Proctor test: High resistance required, of the motorway pavement type.

Table II.7 below summarises the conditions for each test according to the mould used (standard NF P 94-093):

Table II.7 - *Conditions retained for the choice of mould (NF P 94-093).*

<table>
<tr><td rowspan="5">Proctor test</td><td>Test</td><td>Weight of lady (Kg)</td><td>Height of fall (cm)</td><td>Number of strokes per layer</td><td>Number of layers</td><td>Compaction energy Kj/m^3</td></tr>
<tr><td rowspan="2">Normal 2.49</td><td rowspan="2">30.5</td><td>25 (Proctor mould)</td><td>3</td><td>587</td></tr>
<tr><td>55 (CBR mussel)</td><td>3</td><td>533</td></tr>
<tr><td rowspan="2">Modifies 4.54</td><td rowspan="2">45.7</td><td>25 (Proctor mould)</td><td>5</td><td>2680</td></tr>
<tr><td>55 (CBR mussel)</td><td>5</td><td>2435</td></tr>
</table>

11.2.1.5.3 Equipment used

CBR mould (possibly Proctor) ;

Dame Proctor normal or modified ;

Levelling rule ;

Spacer disc ;

Homogenisation tanks for material preparation ;

5 and 20 mm sieves (control and scaling of the sample if necessary);

Trowel, spatula, brush, etc ;

Graduated cylinder 150 ml approx;

Small containers (water content measurements) ;

Weighing capacity 20 kg, accuracy ± 5 g ;

200 g precision balance, accuracy ± 0,1 g ;

Oven 105°C ± 5° C ;

Oil cruet.

11.2.1.5.4 Operating mode

a)- Preparation of samples for testing

Quantities to be sampled

The curve will require at least five "5" trials [1 point (w; Yd) per trial]. Six "6" tests are preferable. For the six "6" measurement points, 15 kg should be taken for the "PROCTOR mould" and 33 kg for the "C.B.R mould".

Checking the sample for test feasibility

If D > 20 mm, the soil must be sieved to 20 mm and the reject weighed. In this case, there are two possible cases:

• If the refusal is < 25%, the test must be carried out in the CBR mould, but without incorporating the refusal (sample cut to 20 mm);

• If rejection is > 25%, the PROCTOR test should not be carried out (hazardous compaction).

Preparation of the sample

• Crush the clods by hand or with a mixer, but not the stones, and carefully homogenise the material (its water content must be homogeneous);

• Air or oven dry the material (3 to 5 hours at 60°C), to facilitate sieving and to start the test with a moisture content lower than the Proctor optimum moisture content (the test is carried out with increasing moisture content);

• Trim the sample to 20 mm (if necessary).

Determining the initial water content

• Experience shows that it's good to have around a 2% difference in water content between each point (harmonious curve). 4% is a maximum.

• It is advisable to start testing at a water content (w) that is about 4 or 5% below wopt (wopt is generally between 10 and 14%).

b)- Preparing the equipment

The choice of mould depends on the size "D" of the coarse grains in the soil and in this case there are two possible cases:

If D < 5 mm (and only in this case), the Proctor mould is authorised, but the CBR mould is recommended;

If 5< D < 20 mm, use the CBR mould (keep the floor intact with all its components);

If D > 20 mm, but refusal < 25 %, the test is carried out in the C.B.R. mould (soil scratched to 20 mm);

Reminder: If D > 20 mm, but refusals > 25%, the Proctor test cannot be carried out.

c) - Performing the test

For the Normal Proctor test, filling is carried out in three "3" layers, whereas for the Modified Proctor test, filling is carried out in five "5" layers. The entire surface must be compacted for each layer as shown in Figure II.2. For the correct execution of the test, the following steps must be followed:

Assemble the mould + base + spacer disc (if C.B.R mould) + paper disc at the bottom of the mould (facilitates demoulding) then weigh the whole assembly "i.e. P1";

Adapt the socket and introduce the 1ere layer and compact it. Place the mould on a concrete base weighing at least 100kg, or on a 25cm thick concrete floor, so that all the energy applied is applied to the sample. Tips: scratch the compacted surface to improve the bond with the next layer;

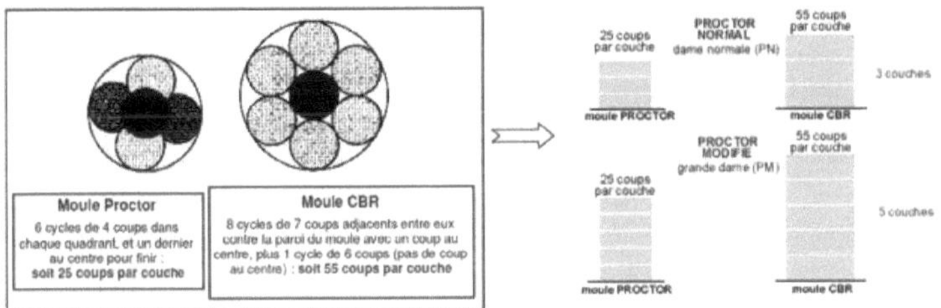

Figure II.2 - Distribution of blows generated on the surface of each layer according to the type of mould used.

Repeat the operation for each layer (three "3" for Normal compaction energy; five "5" for Modified P.) ;

Once the last layer has been compacted, remove the top layer. The compacted soil should protrude from the mould by about 1 cm. If not, repeat the test;

Care should be taken during levelling not to create any holes in the levelled surface;

Weigh the "just levelled" or P2 assembly (weight of the mould + base + spacer disc (if C.B.R. mould) + paper disc at the bottom of the mould (facilitates demoulding) + floor after levelling);

Remove the base (and spacer disc if necessary) and take two "2" samples from the sample, one at the top and one at the bottom; determine the water content w; take the average of the two values obtained;

Increase the water content w of your starting sample by 2% and repeat the test five "5" to six "6" times, after cleaning your mould thoroughly each time.

11.2.1.5.5   Expressing and interpreting results

The dry density « γd » of the sample studied is then calculated by Eq. II.12 below:

$$\gamma_d = \frac{\gamma_h}{1 + \omega} \qquad (kN/m^3) \qquad (Eq.\ II.12)$$

Or :

Yh: Wet density ;

w: Water content.

The wet density $\langle\!\langle\, \gamma h\,\rangle\!\rangle$ of the sample studied is then calculated by Eq. II.13 below:

$$\gamma_h = \frac{P_2 - P_1}{V_{moule}} \qquad (kN/m^3) \qquad (Eq.\ II.13)$$

Or :

P_1: Weight of the "mould + base + spacer disc (if C.B.R mould) + paper disc at the bottom of the mould (facilitates demoulding)" assembly;

P_2: Weight of the "mould + base + spacer disc (if C.B.R mould) + paper disc at the bottom of the mould (facilitates demoulding) + floor after levelling" assembly;

V_{mould}: Volume of the mould used (Proctor mould or C.B.R mould).

Plot the curve Yd = f(w), with the following coordinates for each point: "X-axis: water content (w in %), Y-axis: dry density (Yd in kN/m^3)". From this curve, we can deduce the optimum water content "$_{wopt}$" for which the dry density will be maximum "$_{Ydmax}$".

11.2.2 Hardness tests

In civil engineering, the *"hardness"* of a material is defined as the mechanical resistance that a material opposes to mechanical actions of different kinds. There are a wide variety of possible hardness tests, the most common and familiar being the "*Los Angeles*" and "*Micro Deval*" tests.

11.2.2.1 Los Angeles test drive

The Los Angeles test is used to measure the combined resistance to impact and progressive deterioration due to reciprocal friction of the components of an aggregate. This procedure applies to aggregates used in the construction of pavements and hydraulic concrete.

11.2.2.1.1 Aim of the test

The purpose of the Los Angeles test, in accordance with standard NF P 18-573, is to measure the hardness of a gravel or to measure the resistance of a gravel to fragmentation. It is used to measure the combined resistance to impact and to progressive deterioration caused by reciprocal friction between the components of an aggregate. This method applies to aggregates used in the construction of pavements and hydraulic concrete.

11.2.2.1.2 Principle of the test

The test is used to determine the resistance of a sample of aggregate to fragmentation by impact. The material changes during the test, firstly as a result of the impact of the balls on the aggregate, and secondly as a result of the elements rubbing against each other on the cylinder of the Los Angeles machine. In addition, the test consists of measuring the quantity of elements smaller than 1.6 mm produced by subjecting the material to standard ball impacts.

11.2.2.1.3 Equipment used

A scale accurate to the gram, with a capacity of at least 10 kg;

Sieves;

A "Los Angeles" machine with :

- A hollow steel cylinder 12 mm ± 0.5 mm thick, closed at both ends, with an internal diameter of 711 mm ± 1 mm and an internal length of 508 mm ± 1 mm. The cylinder is supported by two horizontal pins fixed to its two side walls;
- A 150 mm wide opening running the length of the cylinder allows the sample to be inserted;
- The load consists of spherical balls approximately 47 mm in diameter and weighing 420 and 445 g;
- A motor of at least 0.75 kW, ensuring that the machine drum rotates at a regular speed of between 30 and 33 rpm;
- A bin to collect the materials after testing;

• A rotary-type rev counter, which automatically stops the engine at the required number of revolutions.

11.2.2.1.4 Operating mode

The granularity of the material to be tested is selected from the six granular classes (4-6.3 mm; 6.3-10 mm; 10-14 mm; 10-25 mm; 16-31.5 mm and 25-50 mm) of the granularity of the material, as it will be used;

The mass of the test sample will be 5000 g ± 5 g ;

Placement of the sample in the machine as well as the ball load for the selected granular class (Table II.8);

The test is started by rotating the machine 500 times at a regular speed of between 30 and 35 rpm;

Remove the aggregate after the test. Collect the granules in a tray placed under the machine, taking care to place the opening just above the tray to avoid losing any granules;

Sieve the material contained in the bin on the 1.6 mm sieve and weigh the reject, i.e. "m_1", the result of the weighing.

Table II.8 - Relative ball load as a function of the granular class chosen.

Granular classes (mm)	Fractions	Number of balls	Total weight of load (g)	Weight of fractions (g)
4 - 6.3	/	7	3080 ± 20	5000 ± 2
6.3 - 10	/	9	3960 ± 25	5000 ± 2
10 - 14	/	11	4840 ± 25	5000 ± 2
10 - 25	10 - 16	11	4840 ± 25	3000
10 - 25	16 - 25	11	4840 ± 25	2000
16 - 31.5	16 - 25	12	5280 ± 25	2000
16 - 31.5	25 - 31.5	12	5280 ± 25	3000
25 - 50	25 - 40	12	5280 ± 25	3000
25 - 50	40 - 50	12	5280 ± 25	2000

11.2.2.1.5 Calculation of the Los Angeles coefficient (LA)

The resistance to fragmentation by impact of the material is called, by definition, the *"Los Angeles coefficient, LA"* which is expressed by the ratio, of the mass of the elements smaller than 1.6mm produced during the test "m", to the mass of the material subjected to the test "M" multiplied by 100. The lower the Los Angeles coefficient "LA", the more resistant the aggregate is to impact fragmentation. The mass of the fraction of material passing the 1.6 mm sieve test "m": m (g) = 5000-m_1.

(%) (Eq. II.14)

11.2.2.1.6 Expressing and interpreting results

The values of the Los Angeles coefficient indicate the nature of the gravel and make it possible to assess its quality for use in concrete, as shown in Table II.9.

Table II.9 - Type of gravel according to LA coefficient.

Los Angles coefficient values	Appreciation
LA < 15	Good to very good
15 < LA < 25	Fair to good
25 < LA < 40	*Low to* medium
LA > 40	Mediocre "poor quality

11.2.2.2 Micro-Deval test

11.2.2.2.1 Aim of the test

The purpose of the *"Micro-Deval"* test is to determine the resistance to wear due to reciprocal friction between the elements of an aggregate. The European standard EN 1097-1 is used to determine the Micro-Deval coefficient.

11.2.2.2.2   Principle of the test

The test consists of measuring (after sieving) the quantity of elements smaller than 1.6 mm produced in the Deval machine by the reciprocal friction and moderate impact of the aggregates.

11.2.2.2.3   Equipment used

The micro-Deval machine consists of the following components:

One to four hollow cylinders, closed at one end, with an internal diameter of 200 mm ± 1 mm and an effective length of 154 mm ± 1 mm for gravel between 4 and 14 mm and 400 mm ± 2 mm for 25-50 mm. Each cylinder allows one test to be carried out;

The abrasive charge consists of 10 mm ± 0.5 mm diameter spherical stainless steel balls;

A motor (approximately 1 kW) must ensure that the cylinders rotate at a regular speed of 100 rpm ± 5 rpm;

A device must be provided to stop the engine automatically at the end of the test;

We also need :

A scale accurate to the gram, with a capacity of at least 10 kg;

Sieves (1.6 mm sieve and sieves to determine granular classes).

11.2.2.2.4   Operating mode

The granularity of the material tested is chosen from the six granular classes (4-6.3 mm; 6.3-10 mm; 10-14 mm; 10-25 mm; 16-31.5 mm and 25-50 mm) of the granularity of the material as it will be used. For tests carried out on chippings between 4 and 14 mm an abrasive load is used;

The mass of the test sample will be 500 g ± 2 g for 4-14 mm chippings and 10 kg ± 20 g for 25-50 mm aggregates;

Placement of the sample in the machine together with the abrasive load (Table II.10) which is set according to the table for chippings of 4-14 mm and 10 kg of material for aggregates between 25 and 50 mm (without the abrasive load);

Table II.10 - *Abrasive load as a function of the granular class chosen.*

Granular class (mm)	Abrasive load (g)
4 - 6.3	2000 - 5
6.3 - 10	4000 - 5
10 - 14	5000 - 5

Add 2.5 L of water for chippings between 4 and 14 mm and 2.0 L of water for chippings between 25 and 50 mm;

Rotate the cylinders at a speed of 100 rpm ± 5 rpm for :

- 2 h or 12,000 rpm for chippings between 4 and 14 mm ;
- 2 h 20 min or 14,000 rotations for chippings between 25 and 50 mm.

Collect the aggregate and the abrasive charge (for chippings between 4 and 14 mm) in a bin, taking care to avoid any loss of elements;

Sieve the material contained in the bin on the 1.6 mm sieve;

Wash the assembly under a jet of water (remove the abrasive load using a magnet, for example, for gravel between 4 and 14 mm);

Dry the 1.6 mm reject in an oven at 105 °C to constant mass;

Weigh this refusal, i.e. "m1" is the result of the weighing.

11.2.2.2.5   Expressing results

The wear resistance of the aggregate is called, by definition, *"micro-Deval coefficient "MD""* which is expressed by the ratio of the mass of the elements smaller than 1.6 mm produced during the test "m", to the mass of the material subjected to the test "M" multiplied by 100 (Eq. II.15).

(%) (Eq. II.15)

Note: The mass of the fraction of material passing the 1.6 mm sieve test "m":

m (g) = 500- $m1$ for chippings between 4 and 14 mm ;

m (g) = 10000- *ml* for chippings between 25 and 50 mm.

11.2.2.2.6 Interpretation of results

The Micro-Deval coefficient values indicate the nature of the gravel and make it possible to assess its quality for use in concrete, as shown in the following table (Table II.11).

Table II.11 - Type of gravel according to the coefficientMD.

Micro Deval coefficient values in the presence of water	Appreciation
< 10	Very good to good
10 a 20	Good to average
20 a 35	Medium to low
> 35	Mediocre

11.2.3 Load-bearing capacity tests
II.2.3.1 C.B.R. (California Bearing Ratio) test
II.2.3.1.1 Aim of the test

The C.B.R test is a bearing capacity test (the ability of materials to support loads) for embankments and compacted subgrades of road structures. The aim is to experimentally determine load-bearing indices (I.P.I, C.B.R) which enable :

establish a soil classification (G.T.R);

assess the trafficability of earthmoving machinery (I.P.I);

Determine the thickness of pavements (C.B.R increases ^ thickness decreases) or embankments.

Precisely, we measure 3 types of index based on fixed goals:

The Immediate Bearing Index (IBI): This characterises the suitability of the ground to allow the movement of site machinery directly on its surface during the works (H=0 ^ no overloads "S");

Immediate CBR index (ICBR immediat): This characterises the bearing capacity of a compacted subsoil (or pavement component) at different water contents.

The CBR index (ICBR): after immersion: It characterises the bearing capacity of a compacted support soil (or pavement component) with different water contents and subjected to variations in water regime.

II.2.3.1.2 Principle of the test

The load carried by the tyre on the road surface punctures the subgrade. This pointing is smaller the greater the thickness of the road surface (Figure II.3).

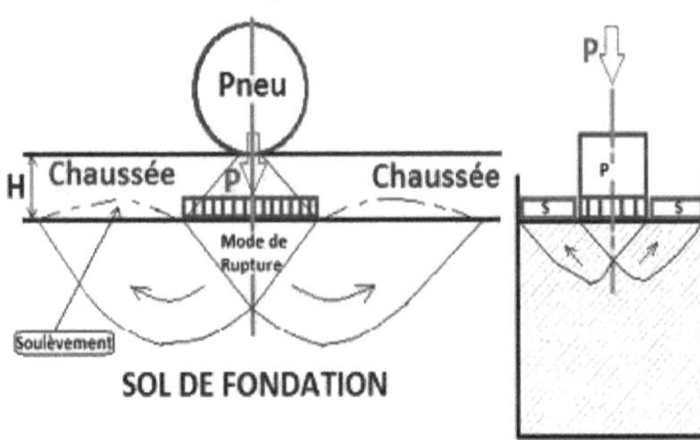

P: pressure applied by the tyre on the subsoil ;

S: surcharge simulating the action of the pavement which opposes the vertical displacement of the subgrade during tyre action.

Figure II.3 - CBR test simulating the pointing phenomenon in a road chase.

This phenomenon can be reproduced by compacting the material under Proctor test conditions in a CBR mould and then measuring the forces to be applied to a cylindrical point to make it penetrate a test tube of this material at constant speed.

It should be noted that the Proctor test must be carried out simultaneously with the CBR test. The thickness of a carriageway depends on the anticipated traffic and axle loads, the underlying soil, and the future water conditions to which the road will be subjected. The expected water conditions during the life of the structure are applied:

Immersion" test for 4 days in water;

Test without *"immersion"*: immediate test.

A load close to that which will be the service load is then applied and the material is punched under specified conditions (constant and specified speed) while the *"forces (F)"* and *"displacements (h)"* are measured, resulting in the following test curve (Figure II.4). Knowing that: "P =F/S" and "S" is the surface area of the point.

Note:

• Standard NF P 94-078 gives the IPI index as a function of applied loads and not stresses;

• These results are then compared with those obtained on a reference soil (crushed waste) (ETALON curve).

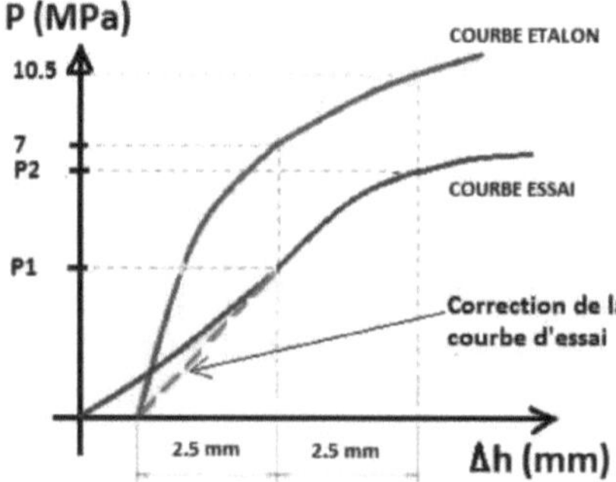

Figure II.4 - Contrast-deformation curve.

11.2.3.1.3 Equipment used

The material at your disposal is a mixture [Clay (15%) + Sand 0/5 (55%) + Gravel 8/12 (30%)] dried in the air or in an oven (the mixture at your disposal must be homogenised);

CBR mould ;

A scale.

II.2.3.1.4 Method of operation

Before inserting the material into the mould :

Secure the base plate to the CBR mould (Figure II. 5);

Place a sheet of filter paper at the bottom of the mould;

Weigh the *"mould + base plate"* assembly when empty;

Determine the volume that will be occupied by the soil sample once compacted;

Fix the sill.

Note: in the case of an IPI test, the spacer disc is not used.

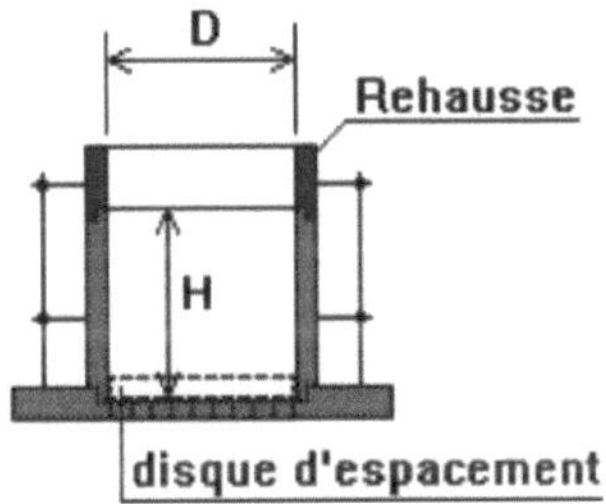

Figure II. 5 - Preparation of the mould.

a)- Manufacturing method

Divide your тёlanдe into equal parts of about 7 kg ;

Spread and moisten each part of your sample to the water content at which you want to carry out the test, then mix by hand to make the mixture as homogeneous as possible;

Note: The approximate quantities of materials to be introduced per layer are as follows (Table II.12):

Table II.12 - Approximate quantities of materials to be introduced per layer.

Mould	Modified Proctor Test (PM), (5 layers)
Proctor	400 g
CBR	1400 g

The corresponding quantity of material is introduced into the CBR mould and compacted in accordance with the conditions of the modified Proctor test (see page 11 of standard NF P 94-093);

Remove the socket and carefully flatten the test tube (from the centre outwards);

Weigh the "*mould + base plate + soil test tube*" assembly to the nearest gram;

Remove the base plate, turn the mould upside down and reattach the base plate;

Remove the sheet of filter paper;

The test piece is then ready for the punching test (Figure IV.6).

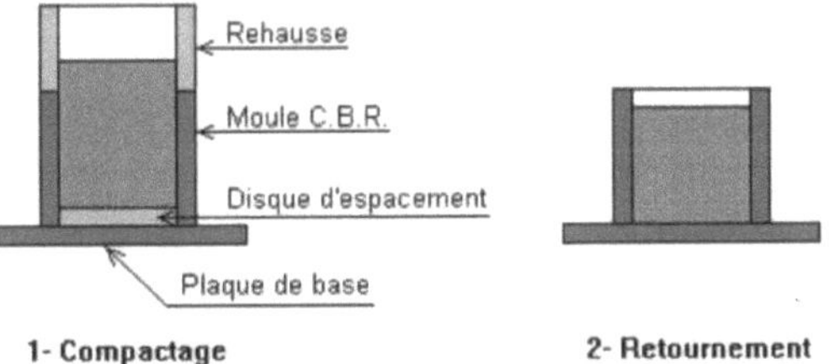

Figure II. 6 - Test tube after preparation.

b)- Determination of the immediate bearing index (IPI)

Place the "base plate, CBR mould and test piece" assembly on the press, centred in relation to the piston;

Before squeezing, move the top of the test tube towards the piston until it is flush with the material *(stop as soon as the needle on the ring moves slightly)*;

Adjust the zero setting of the dynamometer and of the comparator measuring the penetration of the fist;

Fist pumping at constant speed;

Note the punching forces corresponding to indentations of 0.625, 1.25, 2, 2.5, 5, 7.5 and 10 mm and stop punching at this value (Table II.13);

Table II.13 - Punching results corresponding to the different depressions.

t (mn)	0.5	1	1.5	2	4	6	8

Ah (mm)	0.625	1.25	2	2.5	5	7.5	10
F (kN)							
6 (MPa)							

Measure the water content in the vicinity of the punctured area and immediately after the test (at least 2 samples to the left and right);

Carry out at least 4 IPI tests at w = 0% then 4% then 8% and 12%.

c) - Measurement analysis

Plot the punching values measured for the planned indentations on an *effort-deformation* graph (**Note: If the curve shows an upward concavity at start-up, the origin of the indentation scale should be corrected) (Figure II.7);

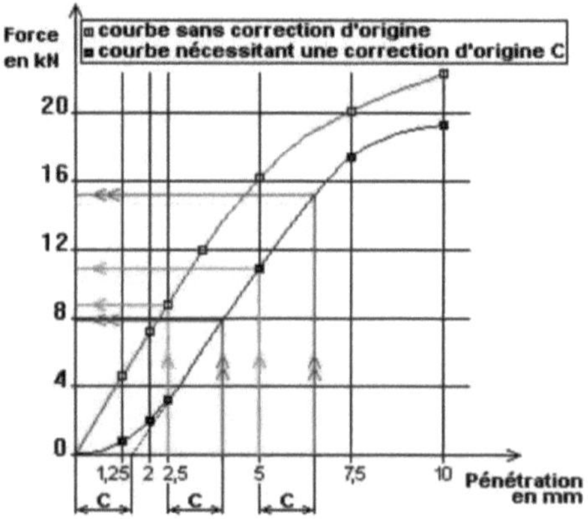

Figure II. 7 - Effort-training curve.

Determine the IPI index in accordance with the standard;

Determine the compaction water content and the dry density p_a (in t/m^3).

11.2.3.1.5 Expressing and interpreting results

We plot the Proctor curve pd = f(w%) ? Compare the two curves and calculate the Sr (%) for each test;

The curve IPI = f(w%) is plotted;

The two saturation curves Sr = 100% and Sr = 80% are plotted in accordance with the standard (Figure II.8).

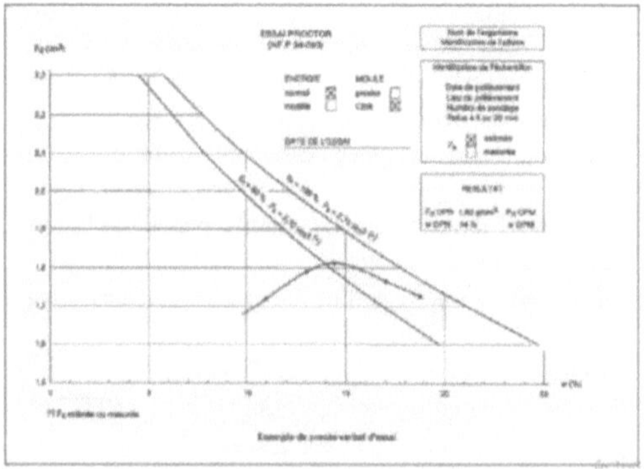

FigureII.8 - Example of a test report.

We determine c_{rpt} and $_{pdmax}$ (from the Proctor test) and also $w\,"_{pi}$ and $_{IPImax}$ (from the CBR test);

The water sensitivity of the material is assessed;

Classification of the soil studied according to the GTR.

If this material is to be used as a road embankment, its ability to withstand the traffic of construction machinery must be assessed according to the following formula (Eq. II.16) :

$$I_{CBR} = Max\left(\frac{P_1}{7}\ ;\ \frac{P_2}{10.5}\right)*100 \qquad (\%) \qquad (Eq.\ II.16)$$

Or :

I_{CBR}: C.B.R index in (%) ;

P_i: pressure corresponding to pressing 2.5 cm into the test piece in (kN) ;

P_2: pressure corresponding to a 5 cm indentation in the test specimen in (kN).

II.2.3.2 Plate test

The earthworks and drainage platforms concerned by this test are road, rail and airport infrastructure constructions, etc. To carry out the test, a reaction mass greater than eight "8" tonnes is required. The test is carried out using two successive loading cycles. The purpose of the "EV2" plate test is to measure the way in which a floor or platform deforms under the application of a heavy load. This method is standardised and refers to standard NF P 94-117-1. For this test to be carried out, it must be ensured that the largest diameter of the aggregates making up the platform does not exceed 200 mm.

11.2.3.2.1 Aim of the test

The plate test is used to calculate the *"modulus under static plate loading, EV2"* of a platform. The plate test is used to assess the deformability of a soil due to the effect of settlement under the loaded plate.

11.2.3.2.2 Principle of the test

The principle of the test is simple. The ground deformation is measured not at the point where it occurs (the actual deformation zone), but at a remote measurement point. To do this, we use a hinged straight beam that forms a line of deformation and pivots around a fixed axis. When the device is unloaded, i.e. the area of soil to be tested is not subjected to any load (Figure II.9), the strain transfer line is confused with the reference line. This is normal. With no load applied to the plate, the deformation of the soil beneath the load-applying plate is zero, as shown in the diagram below.

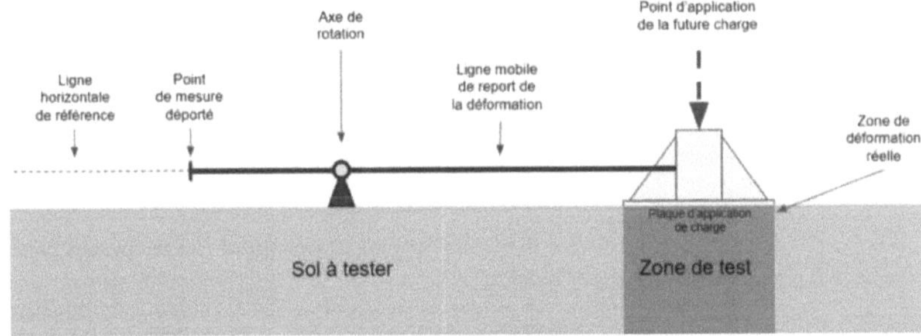

Figure II.9 - No-load measurement device.

When the device is loaded (Figure II.10), i.e. a heavy load is applied to the load plate (1), the ground sinks beneath the plate (2). The moving line (3) then rotates about its axis, and the free end at the measuring point (4) moves upwards. The amount of lift can be measured and the amount of sinkage under the plate deduced, as shown in the diagram below.

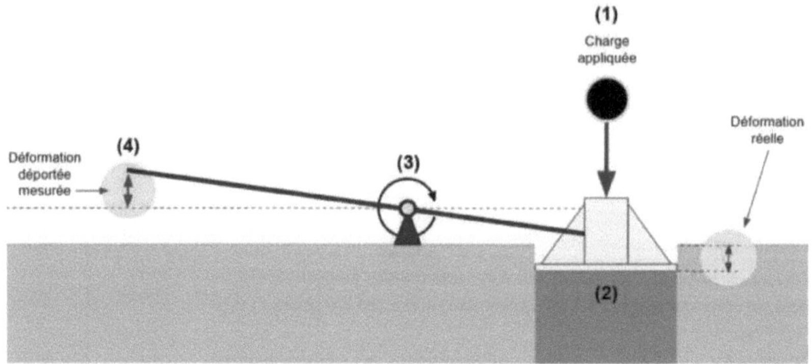

Figure II.10 - Measurement device under load.

11.2.3.2.3 Equipment used

The Benkelman beam is composed of (Figure II.11): A *"jack"*, which is supported under the reaction mass (loaded truck). This jack is equipped with a hydraulic pump which is used to load the system;

A *pressure gauge* is usually built into the pump to check the load being applied;

A *"plate"*, used to apply the load of the reaction mass to a known, uniform surface

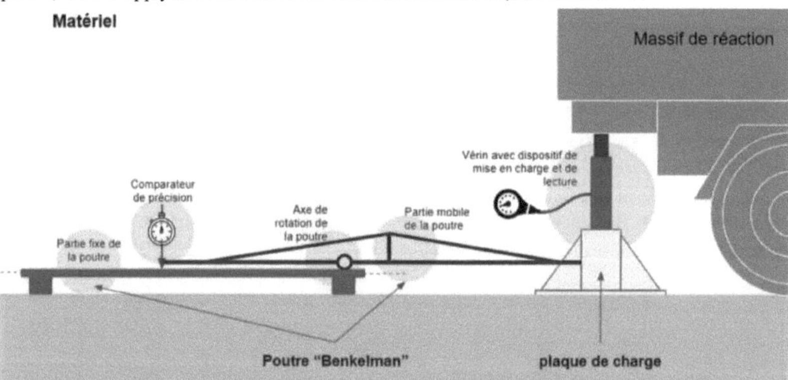

Figure II.11 - Measuring device consisting of a Benkelman beam and a reaction block.

A *"point"* inside the plate is used to check that it is not the plate that is deforming;

A *"moving part"*, which is the link between the *"plate"* and the *"comparator"*. As the plate is pressed in, the beam will rotate around an *"axis of rotation"*;

A *"fixed part"*, supported on the platform, which acts as a support for the *"comparator"*. This part is connected to the *"moving part"* via a *"rotation axis"*.

A *"precision comparator"*, placed at the end of the beam, is used to measure the deflection of the ground under the thrust of the jack. In other words, the comparator measures the sinking (variation in distance) of the ground under the effect of the applied load.

11.2.3.2.4 Operating mode

To carry out this test, two loading cycles are carried out at constant speed (80 daN/s) on a rigid plate 60cm in diameter. The results are calculated on site, then taken into an office to plot a graph of the test. Before the plate is installed, a thin layer of sand is spread underneath the plate. The sand ensures that the load is applied to the entire surface of the plate. The 200 kN ram, lifting the reaction block (6x4), performs two successive loading cycles:

46

1er loading cycle (F=7068 daN): *The* pressure is increased from 0 to 0.25 MPa, then maintained until the deformation stabilises (<0.02mm/15 sec.). We then measure the indentation Z0 in mm and unload (we lower the pressure to zero "0").

2ème cycle de loading (F=5645 daN): Load from 0 to 0.20 MPa then wait for the deformation to stabilise (< 0.02 mm/15 sec.) Measure the indentation "Z2" in mm then unload (reduce the pressure to zero "0"). The value taken is that of the second load "Z2".

11.2.3.2.5   Expressing results

The two loading cycles applied to the platform studied enable the pressure/deformation curve to be plotted (Figure II.12).

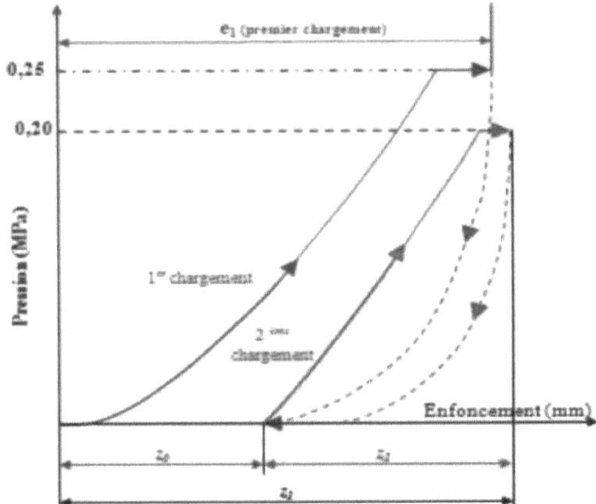

Figure II.12 - Plate deflection as a function of the two pressures applied.

The modulus of deformation at the platform plate at the point of auscultation is determined from the following Boussinesq formula (Eq. II.17):

$$EV_2 = \frac{\pi}{4} * (1 - v^2) * \frac{P * d}{Z_2} \qquad \text{(MPa)} \qquad \text{(Eq. II.17)}$$

Or :

d : Plate diameter in (mm) ;

P: Average effective pressure applied to the ground in (MPa) ;

EV_2 : Static modulus of deformation in (MPa) ;

u : Fish coefficient (without unit) ;

Z2: Plate depression caused by 2eme loading in (mm).

If we equate the expression $(1 - v^2)$ with the value of 1, we obtain approximately the following formula (Eq. II.18):

$$EV_2 = \frac{90}{Z_2} \qquad \text{(MPa)} \qquad \text{(Eq. II.18)}$$

According to LCTP (1973), for the first loading at *"0.25 MPa"*, the value of the modulus *"EVi"* characterises the deformability of the embankment in its state of compaction. Assuming a value for the fish coefficient of *"v = 0.25"*, its expression is given by the following formula (Eq. II. 19):

$$EV_1 = \frac{112.5}{e_1} \qquad \text{(MPa)} \qquad \text{(Eq. II.19)}$$

II.2.3.2.6 Interpretation of results

The *"EV2"* module makes it possible to assess the evolution of deformability during successive loading. In fact, according to the GTR (1992), the minimum bearing capacity of the subgrade before the application of subgrades and pavements is given in Table II.14.

Table II.14 - *Minimum bearing capacity of the load-bearing soil as a function of "EV2" (GTR 1992).*

Type of layer to be applied to the substrate	EV2 (MPa)
Shaped layer in milked materials	> 35
Sub-base in untreated material	> 15 a 20
Pavement layers	> 50

However, a low *"EVi"* value may be due either to insufficient compaction or to compacted material with a very high water content. The ratio *"k = EV2/EV1"* is used to assess the quality of compaction. The lower the *"k = EV2/EV1"* ratio, the better the compaction (Table II.15).

Table II.15 - *Characteristics of foundation support materials (LCPC-COPREC 1980).*

Class of materials	EV2 (MPa)	EV2/EV1
D	> 100	< 2.5
A and B	> 50	< 2

It should be noted that the ratio *"k = EV2 / EVi" is* used to assess the quality of compaction as follows:
If *"EV2 /EVi < 2"*, compaction is good;
If *"EV2 /EVi < 1.2"*, compaction is very good.
A Young's modulus value that can be used directly in a rational sizing scheme can be deduced from the selected CBR value. Several approaches have been proposed:
Geoffroy and Bachelet formula (Eq. II.20) :

$$E = 6.5 * CBR^{0.65} \qquad (GPa) \qquad (Eq.\ II.20)$$

Formula proposed by SHELL (Eq. II.21) :

$$E = 10 * CBR \qquad (GPa) \qquad (Eq.\ II.21)$$

Formula used by LCPC (Eq. II.22) :

$$E = 5 * CBR \qquad (GPa) \qquad (Eq.\ II.22)$$

II.3 Conclusion

Most of the main identification tests on materials used in road pavement structures have been described in this chapter. In this final chapter, we will present the various formulation tests and the materials used in the formulation of bituminous concrete.

CHAPTER III

FORMULATION OF BITUMINOUS CONCRETE

III.1 Introduction

Road pavements are permanently subjected to mechanical and thermal stresses combined with chemical phenomena which, depending on the level of stress, will contribute more or less rapidly to the degradation of the pavement (Perret et al. 2000). These elements, which are responsible for degradation, originate from:

Traffic stresses: mainly the dynamic effects of heavy goods vehicles passing overhead, the static effects of traffic slowing down and tyre/road surface friction;

Climatic stresses: leading to temperature variations within the asphalt mixes. These variations are either short term (daily) or long term (seasonal);

Chemical phenomena: due to the natural oxidation of hydrocarbon binders, the action of de-icing salts from winter maintenance, and damage to the surface of the pavement from solar radiation.

These various actions, acting simultaneously on the bituminous surface, lead to the deterioration commonly observed (VSS SN 640 925a 1997), in particular:

Surface cracks in the form of isolated cracks or, in the most serious cases, in the form of generalized cracking;

Permanent deformation (or rutting) resulting from the accumulation of irreversible deformation;

Surface degradation in the form of aggregate polishing, stripping and loss of chippings, peeling and potholes.

As a result, the need to improve or optimise the durability of road pavements, both when designing new pavements and when maintaining existing ones, is a major concern for road network managers. With this in mind, it is essential to assess the long-term performance of asphalt mixes using relevant laboratory tests. The new load configurations of heavy goods vehicles have significantly altered the stresses on road surfaces, leading to premature deterioration of carriageways that were previously behaving normally, and whose materials met the specifications in force. In particular, road surfaces are suffering from the gradual replacement of twin wheels on heavy goods vehicles with super-wide tyres (300 to 500 mm wide), and generally higher inflation pressures. These findings have highlighted the fact that some of the tests currently used are not, or are no longer, suitable for assessing the performance of asphalt mixes and should be replaced.

The components and mixes must therefore be characterised by laboratory tests which will make it possible to assess the long-term performance of bituminous mixes with a view to optimising their service life. The mix design is a decisive factor in obtaining high-performance bituminous mixes. The composition of asphalt mixes has a decisive influence on the durability and performance of pavements.

To date, the optimisation of formulas for bituminous mixes is still based on an empirical approach using traditional tests which often show little correlation with the actual performance of the materials. In order to meet the durability criteria for road surfacings, it will be necessary to select a range of tests that will enable the performance of the bituminous mixes used in pavements to be satisfactorily assessed.

III.2 Bitumen

III.2.1 Definition

Asphalt concrete is a bitumen-rich coating made from a mixture of aggregates (sand, gravel and fines), used mainly for wearing courses, i.e. the top layers of the road surface. Bituminous concrete is classified according to its gradation. They are always laid on a base course of hydrocarbon materials or treated with hydraulic binder, or on a binder course of asphalt mix for thin layers (Figure III.1). More precisely, asphalt concrete is composed of different elements, namely :

Gravel ;

Sand ;

Fillers ;
Bitumen used as a binder.

Bitumen Aggregates Asphalt

Figure III.1 - *Composition of an asphalt mix used mainly for wearing courses.*

111.2.2 Origin and manufacture

All bitumens are products of crude oil, where they are found in solution. They are the result of the elimination of the oils used as solvents by evaporation or distillation of the crude oil. Given that such processes could occur in nature, in underground layers, bitumen comes from two sources: natural or industrial.

Natural origin: World production is very low, not exceeding 200 thousand tonnes;

Industrial origin: This consists of two parts (Figure III.2):

• *Direct distillation"*: Atmospheric distillation: This refining method involves continuously heating the crude oil, which has previously been decanted and desalinated, by passing it through a furnace. This crude, 75

heated to around 340°C, is sent to a fractionation column maintained at atmospheric pressure. The product recovered at the bottom of the tower is the reduced crude;

• *Vacuum distillation"*: At this stage, the reduced crude from atmospheric distillation is heated to around 400°C and then fed into a column under reduced pressure. With this type of unit, all grades of bitumen from 20/30 to 160/220 can be produced directly.

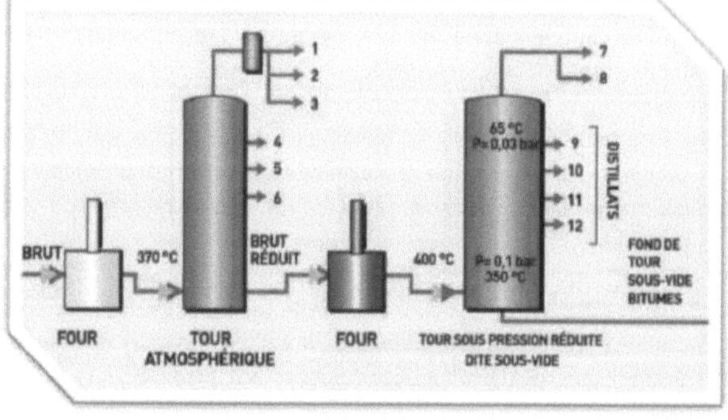

1.	Gaz	5.	Carburateur	9.	Gasoil sous –vide
2.	Essence légère	6.	Gasoil	10.	1ᵉʳ Sous -vide
3.	Essence	7.	Vers éjecteurs de vapeur	11.	2ᵉ Sous -vide
4.	White-spirit	8.	Gasoil entraîné	12.	3ᵉ Sous -vide

Figure III.2 - *Production of bitumen by refining oil.*

111.2.3 Mechanical and rheological properties of bitumen

Bitumen gives the asphalt its flexibility and its ability to withstand certain degradations caused by several factors including traffic, climatic conditions on the site, etc. At high service temperatures, bitumen must remain sufficiently viscous to prevent rutting. Conversely, at low temperatures, the bitumen must retain a certain elasticity to prevent thermal shrinkage cracking and embrittlement of the asphalt. At intermediate temperatures, it must be resistant to fatigue under the effect of repeated traffic. Two tests can be carried out in order to characterise the bitumen intended for the formulation of wearing course mixes, namely :

Needle penetration test (NF T66-004) ;

Ball and ring softening point test (NF T66-008).

111.2.4 Different types of asphalt concrete

There are many different types of *asphalt concrete,* each with its own characteristics and uses. The choice of asphalt concrete depends on the climate of the region, which must be taken into account when selecting the future surfacing because :

The bitumen used can soften if the heat is too strong;

Concrete can harden at too low a temperature;

Rain or heavy snow are also criteria to be taken into account, as BB can freeze and encourage the formation of black ice.

111.2.4.1   Thin asphalt concrete (BBM)

It has a grading of 0/10 or 0/14. This asphalt concrete is easily compacted and is perfectly waterproof. Mainly used for car parks and pavements, its thickness varies from 2.5 to 5 cm.

111.2.4.2   Very thin asphalt concrete (BBTM)

This is undoubtedly the most interesting coating in terms of value for money. It has a very good service life and is easy to lay. Whatever the grading (0/10 or 0/6), BBTM has a thickness of between 1.5 and 3 cm.

111.2.4.3   Ultra-thin asphalt concrete (BBUM)

It is designed for wearing courses, i.e. in direct contact with vehicle tyres, and is ideal for use in car parks, for example. Its thickness varies from 1 to 1.5 cm.

111.2.4.4   Flexible asphalt concrete (BBS)

As its name suggests, this concrete is made from a bitumen that is soft enough to produce a deformable coating. However, it has low resistance to rutting.

111.2.4.5   Draining asphalt concrete "BBDr

The main advantage of this type of asphalt is its excellent grip, whether it's raining or hot. It also reduces rolling noise. So it's perfectly suited to garden paths.

111.2.4.6   Semi-grained asphalt concrete (BBSG)

This is the benchmark hot mix asphalt. It meets a wide range of needs (pavements, driveways, etc.) and is highly suitable for both medium and heavy traffic. Its thickness varies from 3 to 9 cm, depending on the grading.

111.2.4.7   High modulus asphalt concrete (BBME)

This asphalt concrete is a structural asphalt. It has excellent rigidity, a long service life and good resistance to rutting. Its thickness can vary from 4 to 9 cm per layer.

III.2.5 Compaction of asphalt concrete

111.2.5.1   Pre-compacting bituminous mixes

Pre-compaction is achieved by the vertical vibration of the tamper or the screed, or a combination of both.

combination of these two elements. Pavers with a screed or a vibratory tamper can achieve a compaction rate of 80 to 85%, whereas pavers with a screed and a vibratory tamper can achieve a higher compaction rate (depending on the mix composition and the forward speed). The vibratory tamper should be positioned over the full width of the screed, slightly below screed level. If the tamper travels too short, it will be ineffective, but if it travels too long, it can cause surface tearing.

The frequency of vibration should be adapted to the composition of the asphalt mix and the thickness of the layer; if the layer is thin, the frequency should be limited. High, uniform pre-compaction is important for final evenness. The screed must be thoroughly preheated before spreading the mix. This preheating, which must be uniform, takes longer for the surface layers. Local overheating of the screed can cause deformation of the screed, resulting in profile irregularities.

111.2.5.2   Compacting hot mix asphalt

It is necessary to check the flatness and transverse slope of the surface behind the screed. The aim of the compaction phase is to increase the density of the spread asphalt layers in order to improve their resistance, while maintaining the surface characteristics of evenness and adhesion required for safety and comfort. It thus gives the material definitive characteristics that will be directly perceived by users. Good compaction ensures :

Improved creep resistance;

Improved resistance to fatigue and a longer life as a result;

Good surface flatness and adequate roughness.

At present, compaction is carried out in a static mode on our sites, mainly using a pneumatic compactor. The results of experimental trials carried out mainly in the Algerian context have prompted the introduction of dynamic compaction (vibratory compactor). In this phase of the work, the following four important points need to be developed:

The compaction equipment and its characteristics ;

The field of use of compactors ;

Action to be taken on parameters influencing compaction ;

Types of compaction workshop.

111.2.5.3   Mix compaction equipment

The compactor is nothing more and nothing less than a steamroller, essential for roadworks. Its role is to compact and level the asphalt on the ground. Compaction usually begins at the joints and at the edge of the carriageway. Another pass will compact and press the pavement. In most cases, the better the compaction, the better the performance. Road surfaces are basically made up of solid materials such as aggregates and sand, liquid materials such as bitumen and emulsion, and air. Compaction stabilises the soil by bringing the materials together and compacting them, but also by expelling the air. Compaction is therefore a final and delicate operation on the site. It is a decisive stage in terms of the durability of the pavement as a result of the compactness achieved, but also in terms of the surface quality in terms of evenness and texture. The main machines used in the production of asphalt concrete (wearing course) are shown in Figure 111.3.

Pneumatic compactor Asphalt paver feeder

Asphalt paver Smooth roller compactor

Compactors are classified into four types:

Tyre compactors for surfacing materials": these are generally quite mobile (speeds of between 3 and 6 km/h), and are often chosen for sandy or clayey soils, if they can do just about anything: earthworks, pavements, coatings, etc;

Static foot-operated compactors": these are used mainly for large-scale earthworks (speeds ranging from 6 to 12 km/h);

Static compactors with smooth cylinders": these are reserved for compacting asphalt of less than 4 cm, i.e. very fine asphalt;

Vibratory compactors with smooth rollers or tamping feet: these can be used for most applications, particularly when laying asphalt or floating or thick materials (speeds ranging from 3 to 12 km/h).

The compacted material is brought to a void percentage that allows the desired performance to be obtained by using one of the following compaction methods or a combination of some of them.

Compression compaction" defined by the effect of the contact pressure of the wheel with the surface material;

Compaction by tamping" is defined by the effect of the wheel load in the lower part of the layer;

Vibration compaction" is defined by the effect of vibration of the cylinder, ensuring compaction by rearrangement of the grains.

III.3 Methods for formulating bituminous concrete

To date, no universal method has been developed that is totally free of empiricism, which has led researchers to develop several methods of formulation, from which enormous progress has already been made.

111.3.1 Main objective of the formulation

The main objective of the mix design is to determine an optimum composition of *"aggregates"*, *"binders"* and *"voids"* that will enable the following *"performances"* to be achieved, namely :

The complex modulus E^* ;

Resistance to mechanical fatigue ;

Resistance to permanent deformation ;

Resistance to thermal stress ;

Resistance to hydraulic shrinkage cracking.

Other non-mechanical performances must be considered in order to characterise the performance of pavements:

Adhesion ;

Resistance to tearing ;

Ageing ;

Hydraulic susceptibility.

111.3.2 Hveem method

The main concepts behind this method were set out by *Francis N. Hveem*, an engineer with the *Californian Department of Transportation (CDT)*, in the 1930s. This method subsequently underwent various improvements to become the official CDT method (Asphalt Institute 1997). The formulation procedure can be defined in several stages:

111.3.2.1   Choice of materials

This choice must be in accordance with the project specifications. The materials must meet the physical and chemical properties set out in the specifications.

111.3.2.2   Choosing the particle size distribution curve

The combination of the different aggregate sizes should make it possible to obtain a grading curve that is as close as possible to the reference curve set out in the specifications.

111.3.2.3   Determining the approximate binder content

This estimate is based on two tests specific to this method: the *Centrifuge Kerosene Equivalent (CKE)* and the *Surface Capacity*. Based on these tests and the actual density of the fines and stones, specific charts can be used to estimate the *"optimum binder content"*.

111.3.2.4   Preparing samples

The samples are manufactured according to a standard procedure and in standard moulds. Compaction is carried out using a mechanical compactor with a method that is also standardised. One sample should be prepared with the binder content obtained previously, two with lower binder contents *"-0.5% and -1.0%"* and one with a higher binder content *"+0.5%"*.

111.3.2.5   Stability and penetrant testing

Once compacted, the samples are subjected to these two tests. The equipment used for the *"stability test"* is specific to the *"Hveem method"*; minimum stability values are set according to the traffic. The *dye penetration test* is more qualitative.

111.3.2.6   Choosing the optimum binder content

The optimum binder content is the maximum binder content for which the sample satisfies the minimum stability conditions, does not show significant bleeding and the binder content is at least 4%. If this value is that of the maximum binder content prepared (estimate +0.5%) then an additional sample with a binder content +0.5% higher must be prepared and the procedure repeated.

111.3.3 Marshall Mix Design

The first concepts of this method were developed by Bruce Marshall in the late 1930s, then revised and improved by the *U.S. Army*. This method, recommended by the VSS standards in Switzerland, aims to select the binder content, for a certain density of mix, which satisfies minimum stability and creep within an acceptance range (Asphalt Institute 1997). The formulation procedure can be summarised in six distinct stages:

111.3.3.1   Choice of aggregates

Aggregates are chosen according to their physical characteristics (hardness, cleanliness, shape, etc.). Once this choice has been made, their particle size and density are determined, and then the various aggregates required to obtain the reference particle size curve are selected.

111.3.3.2   The choice of binder

As this method does not have a standardised selection and evaluation procedure, the choice is left to the engineer, who must carry out the tests he deems necessary.

111.3.3.3   Preparing samples

The samples are made in standard moulds. Typically five *"5"* mixes with different binder contents are prepared, and for each mix three samples are taken. The samples are then compacted with a hammer to standardised dimensions and according to precise rules.

111.3.3.4   Determining stability and creep

Once compacted, the samples are subjected to a stability and creep test. Stability is the maximum force that the sample can withstand and creep is the resulting plastic deformation. These two values are, in a way, measurements that allow us to predict the performance of the coating.

111.3.3.5   Calculating density and voids

The density and voids (voids in the mixture, voids in the mineral skeleton, voids filled by the bitumen) are used to characterise the mixture.

111.3.3.6   Choosing the optimum binder content

This choice depends on the combination of results for stability and creep, voids and density. Thus six *"6"* graphs representing the evolution of the percentage of voids, density, creep, stability, voids in the mineral skeleton *"VMA"* and voids filled by bitumen *"VFA"* *as a* function of the binder content are plotted. The choice of the percentage of voids in the mix makes it possible, on the one hand, to obtain the optimum binder content and, on the other hand, to check whether this binder content meets the requirements for the other parameters. These two actions are carried out graphically using the curves obtained from the tests on the samples.

111.3.4 SUPERPAVE Mix Design" American method

This formulation method was developed in the United States to replace the Marshall method. It is widely used in the field. In 2000, 62% of the total production in tonnes of bituminous surfacing was produced using this method. As part of the *Strategic Highway Research* Program *(SHRP)*, whose objectives were to improve the choice of materials and the formulation of bituminous mixes, a new formulation method was developed in the early 1990s: the SUPERPAVE *Superior Performing Asphalt Pavement* method (WTFTCR 2001). The *SUPERPAVE* mix design method is based on the concept of workability during laying and the performance of the asphalt over time. It can be divided into four stages (Asphalt Institute 2001):

111.3.4.1 Choice of aggregates

This choice is based on three different criteria. The granulometric curve, which must lie between two limits and pass through fixed points. *Consensus properties*: angularity, shape and sand equivalent. Source *properties*: hardness, noise and cleanliness.

111.3.4.2 Choice of binder

This choice depends not only on the physical characteristics of the binder (penetration, viscosity, etc.), but also on climatic conditions and the type of traffic. There is a *performance grading* system for binders
(PG) system" which is a function of the maximum and minimum temperature of the pavement and the traffic conditions. The determination of these three parameters, combined with the minimum reliability of the *"reliability design"* result, makes it possible to define the minimum *grade of* binder to be used.

111.3.4.3 Choosing the optimum binder content

Samples were produced with four different binder contents (variation of ±0.5%) and subjected to compaction in the *Superpave "Gyratory Compactor: PCG"* or *"gyratory shear press"*. The maximum theoretical density of the mix is plotted as a function of the number of gyrations for the four samples. We then graphically determine the binder content that satisfies the desired percentage of voids and the number of gyrations required to obtain it. The number of gyrations is defined by the traffic conditions.

111.3.4.4 Performance testing

This last stage, still under development, will standardise the tests to be carried out to determine the mechanical characteristics of the mixes, namely the *"dynamic modulus"*, the *"flow time"* and the *"flow number"* (Bonaquist et al. 2003).

111.3.5 Methode Fran^aise

This method (used in Algeria) is based on two main principles. The first is the determination of the minimum quantity of binder as a function of the granulometry of the mix. The second is the use of a gyratory shear press to estimate the behaviour of the mix during compaction.

III.3.5.1 Minimum quantity of bituminous binder

In the approach that was codified in France in the 1950s, for a given aggregate composition, a minimum quantity of bituminous binder is defined to ensure good mix durability by the concept of *"richness modulus: K"* $(2 \leq k \leq 2.6$. *for GB and* $3.3 \leq k \leq 3.9$. *for BB)*. This quantity, which is proportional to a conventional thickness of the binder film coating the aggregates, is given by the following expression which links *"K"* to the binder content *"TL"* and to the *"conventional specific surface area of the aggregates:* Σ *»* (Eq. III. 1):

$$\text{TL} = \text{K} * \alpha \sqrt[5]{\Sigma} \qquad (\%) \qquad (\text{Eq. III.1})$$

Or :

a: Corrective coefficient relating to the density of the aggregates calculated using the expression below (MVR$_g$ is the actual density of the aggregates) (Eq. III.2) :

$$\alpha = \frac{2.65}{\text{MVR}_g} \qquad (-) \qquad (\text{Eq. III.2})$$

Σ : The conventional specific surface area is calculated using the expression (Eq. III.3):

$$100\Sigma = 0.25G + 2.3S + 13s + 135f \qquad (-) \qquad (\text{Eq. III.3})$$

With mass proportions :

G : elements larger than 6.3 mm ;

S : elements between 6.3 mm and 0.315 mm ;

s : elements between 0.315 mm and 0.08 mm ;

f: elements smaller than 0.08 mm.

III.3.5.2 Gyratory shear press test

The composition of the mix is chosen on the basis of previous experience, and its behaviour during compaction is estimated using the *"gyratory shear press"* test. A pre-determined quantity of the hydrocarbon mixture, heated to the temperature normally used to manufacture the mix in a plant, is placed in a cylindrical mould with a diameter of 150 mm or 160 mm. Compaction is achieved by the simultaneous action of :

A fairly low static compression force corresponding to a pressure of 0.6MPa;

A deformation of the test-tube which is required to have its longitudinal axis describe a conical surface of revolution, with vertex *"O"* and vertex angle *"2a"*, while the end surfaces of the test-tube remain substantially horizontal. The angle *"a"* is approximately *"1°"*. It is determined for each type of machine in order to obtain fixed percentages of voids on materials taken as a reference. The speed of rotation has little influence on the result, and is usually taken to be equal to *"30 rpm"*.

The interpretation of the test from the point of view of assessing the behaviour during compaction of the mix is made by considering the void percentage values obtained in general after *"10 gyrations"* and after a *"number of gyrations: Ng"* of [25, 40, 60, 80, 100, 120 or 200] which depends on the type of mix being studied. After *"10 gyrations"*, a minimum value for the percentage of voids is generally specified (of the order of *"14%"* for base course materials) in order to avoid having a mix that is too easy to handle, which would prove difficult to compact without excessive deformation and would lead to a material that would also prove unstable under traffic.

After *"Ng girations"*, a range of values is specified for the percentage of voids (for common bituminous concretes used in surface courses, the percentage of voids must be between 5% and 10%. For mixes intended for base courses, only the maximum value is specified, which can be as high as 11%). The maximum value is intended to ensure the durability of the mix, while the minimum value is intended to avoid excessive compaction, which would encourage the instability of the mix and the development of rutting due to creep under traffic, as well as to ensure the maintenance of a sufficient macro-texture for wearing courses.

III.3.6 Belgian method "CRR formulation

The Belgian *"Centre de Recherche Routiere: CRR"* has also developed a mix design method for bituminous mixes. This is characterised by the fact that it is analytical; the use of a laboratory test is only justified as a means of verifying the values determined by the volume mix design. The mix design procedure is identified in Figure III.4 and can be summarised in three phases (CRR 1987, 1997):

III.3.6.1 Choice and characterisation of materials

As with the other methods, the choice of mix components is of great importance for the result and performance of the coating. It is important to know the physical characteristics of the aggregates (grain size, hardness, cleanliness, etc.) and the binder (penetration, thermal susceptibility, etc.) (Figure III.4).

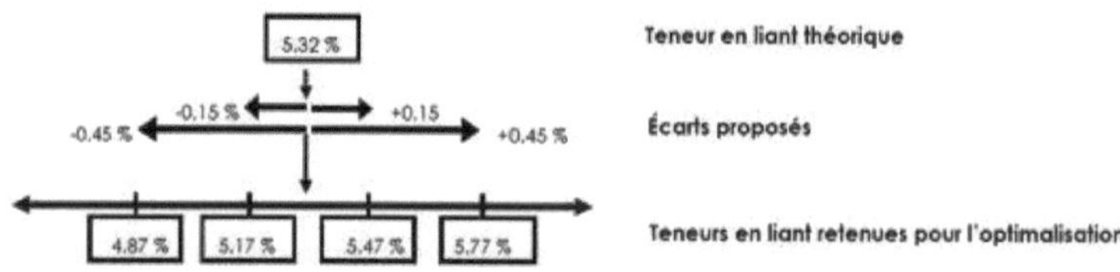

Theoretical binder content
Proposed differences

Figure III.4 - Example of binder content selection for compactibility optimisation.

III.3.6.2 Formulation based on an analytical method

The composition of a mixture must first be designed on the basis of an analytical method that takes account of the important parameters. This phase is integrated into the *"PRADO specific software"*.

III.3.6.3 Verification of formulation results

The use of a laboratory test enables the results of the analytical formulation to be verified and, if necessary, any discrepancies to be corrected by acting either on the composition or on the materials. Only when the test agrees with the results can we move on to the final stage, which is the transition from a volumetric formulation to a mass formulation, using the densities of the components, and the calculation of the mechanical characteristics of the mix thus obtained.

III.3.7 Formulation method used in Algeria

The *"formulation in Algeria"* is based on the *"verification of the characteristics of the components"* as well as the two *"Duriez"* and *"Marshall"* tests according to the granular materials. A formula is chosen which gives a mix with the best *"compactability"* and which could give a better *"stability to the hydrocarbon mix"*. The granular fractions are chosen from the following: *"0/3, 3/8, 8/15"*, the characteristics of the aggregates are represented as follows (Table III.1):

Table III.1 - "0/14" granular fractions of "BB-0/14" asphalt mixes.

Sieve capacity (mm)	BB-0/14
20	-
14	94-100
10	72-84
6.3	50-66
2	28-40
0.08	7-10

By definition, the bitumen content is the mass of binder on the mass of dry aggregates expressed as a percentage, and for this we use the formula *"Eq. V.1"* described in *"paragraph V.3.5"*. Compactness: *C"* is a direct consequence of the formulation, for this calculation it is necessary to know the apparent density of the specimen $\ll \gamma_{app} \gg$, the density of the bitumen $\ll \gamma_b \gg$, the density of each of the aggregates " $\ll \gamma_{G1}, \gamma_{G3}, \gamma_{G2}, ... etc. \gg$, the percentages by weight of each of the constituents in relation to 100 (binder and filler included). The real density *"γ_{rel}"* of the coated material is therefore calculated as follows (Eq. III.4):

$$\gamma_{rel} = \frac{100}{(P_b/\gamma_b) + (P_{G1}/\gamma_{G1}) + (P_{G2}/\gamma_{G2}) + \cdots} \qquad (kN/m^3) \qquad (Eq.\ III.4)$$

Or :

γ_{app} : the apparent density of the test tube ;

γ_b : the density of bitumen ;

$\gamma_{G1}, \gamma_{G2}, \gamma_{G3} \ldots etc.$: the densities of aggregates 1, 2, 3, etc. ;

Pb: the percentage by weight of bitumen ;

P_{G1}, P_{G2}, P_{G3} - etc. the weight percentages of the aggregates.

The volumetric percentage of voids *"VV"* in the test piece is determined by the formula described below (Eq. III.5):

$$V_V = \frac{(\gamma_{rel} - \gamma_{app})}{\gamma_{rel}} * 100 \qquad (\%) \qquad (Eq.\ III.5)$$

The compactness *"C"* of the mix is determined by the following formula (Eq. III.6) :

$$C = 100 - V_v \qquad (\%) \qquad (Eq.\ III.6)$$

III.3.8 Parameters influencing the choice of a formulation

The main characteristics consist in choosing the aggregates, the binder and the additives used to manufacture the coating. This is based on the following considerations:

Traffic; volume, percentage of heavy goods vehicles, axle load ;

Climate: rainfall, frost, temperature, sunshine;

Position of the layer: wearing course, base, bound foundation ;

Layer function: adhesion, permeability, noise, rutting, etc.

III.4 BB formulation stages and procedure

111.4.1 BB" formulation stages

The various stages in formulating bituminous concrete can be summarised in the diagram below (Figure III.5):

Note:

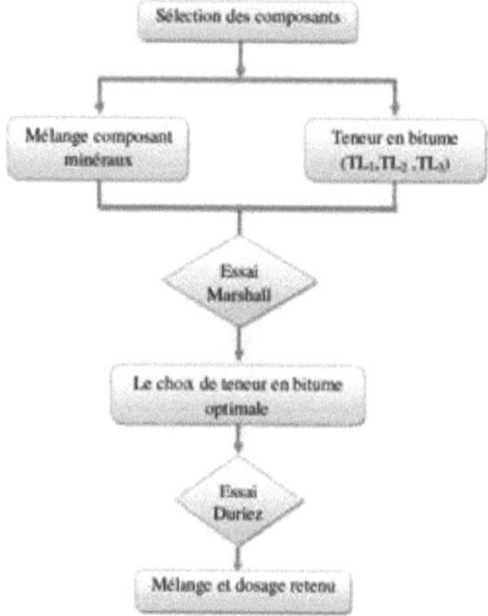

Figure III.5 - Formulation stages.

The characteristics of the aggregates used in the mix design must be determined by the following tests:

Density (NF P18-560) ;

Summary chemical analysis (NF 15-461) ;

Particle size analysis (NF P18-560) ;

10% fine sand equivalent (NF P18-597) ;

Surface cleanliness (NF P18-591) ;

Flattening tests (NF P18-561) ;

Los Angeles test (NF P18-573) ;

Micro-Deval test in the presence of water (MDE) (NF P18-572).

111.4.2 Test equipment and procedures

III.4.2.1 Marshall test

The concept of the Marshall test was developed by "*Bruce Marshall*" in 1948 at the Mississippi State Highway Department, USA. This test enables the resistance of a specimen to deformation under gradual application of a load to be measured in the laboratory at a given temperature and compaction energy, and the deformation undergone by this specimen at the moment of failure under the application of the maximum load, known as *"stability"* and *"Marshall creep"*. These last factors give an indication of the overall quality of the mix, including the choice and dosage of constituents to obtain the best composition or formulation for a mix (stability has a maximum for a certain bitumen content, then it decreases).

III.4.2.1.1 Principle of the test

The Marshall Stability Test is a compression test applied along the generator of a semi-braided cylindrical specimen (Figure III.6). This compression is applied to the specimen after 30 minutes immersion in a water bath at 60°C, and at a speed of 0.85mm/s ± 0.1mm/s.

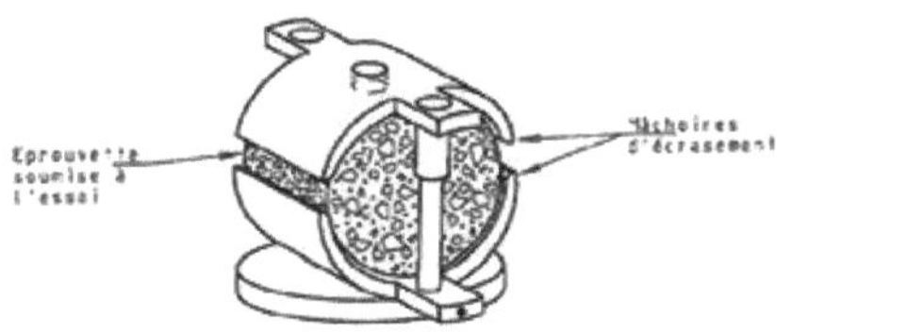

Figure III.6 - Marshall test machine.

III.4.2.1.2 Test procedure

Once the test-tubes have been prepared, one part is used to determine the apparent density and the other to determine the stability and creep. The test specimens are immersed in the thermostatic bath set at 60°C ± 0.5°C with the crushing jaws for 30 min (± 1 min). During this time, the rate control device set for a speed of 0.85 mm/s ± 0.1mm/s is also installed. The specimen is placed in the crushing jaws, and the assembly is carried between the press platens to be subjected to the compression test. These operations must be completed in less than a minute. Breakage occurs when the machine stops, and the figures shown on the machine's display for *"stability"* and *"creep"* are then noted (NF P98-251-2). The same steps are taken for all the specimens produced.

III.4.2.1.3 Expression of results

This test makes it possible to obtain the stress at fracture, which is given by the following equation (Eq. III.7):

$$\sigma_{rupture} = \frac{2P}{\pi h D} \qquad (N/mm^2) \qquad (Eq.\ III.7)$$

Or :
P: breaking load (N)
h : height of test piece (mm)
D : diameter of the test tube (mm)

The strain at fracture during the test at 45°C is also calculated (Eq. III.8):

$$\varepsilon_{rupture}(45°C) = \frac{\Delta D}{D} \qquad (-) \qquad (Eq.\ III.8)$$

Or :

$\Delta D =$ vertical deformation
D = diameter of the test tube

III.4.2.2 Duriez test

The purpose of the *"Duriez"* or *"compression-immersion"* test is to characterise the compressive strength and resistance to stripping by water of coated materials. This test enables the water resistance of a hydrocarbon mixture to be determined at 18°C for a given compaction, based on the compressive strength ratio before and after immersion of the test pieces. The *"Duriez"* test is carried out on the specimens that give the best *"Marshall stability"* corresponding to the *"optimum bitumen"* content.

III.4.2.2.1 Principle of the test

The test specimens required for the test are produced by double-acting static compaction. The test specimens are subjected to the compression test after storage at 18°C under defined conditions: some specimens in air, others immersed for 7 days. The resistance to water is characterised by the ratio of resistances before or after immersion (Figure V.7).

Figure III.7 - Duriez test machine with an example of a made-up test piece.

III.4.2.2.2 Test procedure

D being the day on which the test tubes are made, storage without immersion begins on day *"D + 1"*. The test tubes are stored at 18°C ± 1°C and in an environment with 50% ± 10% relative humidity for 7 days. On day *"D +8"*, the test tubes are subjected to the compression test, whether they have been stored with or without immersion. For each test tube, the time between the exit of the temperature maintenance device and the start of the crushing is less than 2 min. The speed of the press platen is set at 1 mm ± 0.1 mm. The *"simple compressive strength"* is determined from the maximum load at break of the test specimen expressed in *"Kg"*, dividing by 50 gives the *"compressive strength"* expressed in *"Kg/cm² "* which is called *"Duriez stability"* (NF P98-251-1). The same steps are taken for all the specimens produced.

III.4.3 Production of test specimens for the "Marshall" and "Duriez" tests

The correct preparation of the test specimens is an essential step. Asphalt concrete specimens are made according to the type of test, either *"Marshall"* or *"Duriez"*.

111.4.3.1 Preparation of mixtures

Asphalt mixes can be prepared in the laboratory in accordance with standard NFP 98-250-1 for either the *"Marshall"* test or the *"Duriez"* test.

111.4.3.1.1 Preparation of aggregates

Each type of aggregate used in the composition of the hydrocarbon mixture must be sampled in accordance with standard NF P18-553 *"Preparation of a sample for testing"*. The various aggregates are heated in containers in an oven at 160°C.

111.4.3.1.2 Preparation of the binder

A quantity of binder corresponding to the needs of the test is withdrawn without exceeding 100°C. The collected binder is placed in a filled and sealed container. The binder is heated in two stages:

The filled and sealed container is placed in an oven and heated to the reference temperature (generally between 140°C and 180°C);

The container is placed on a hot plate and its contents shaken constantly to homogenise its temperature and maintain it at its reference temperature. This operation should not last longer than 10 minutes.

The reference temperature for the preparation of mixes is defined according to the category of hydrocarbon binder used as follows (NFP 98-250-1): Bitumen 80/100: 140°C ± 5°C, Bitumen 60/70: 150°C ± 5°C, Bitumen 40/50: 160°C ± 5°C and Bitumen 20/30: 180°C ± 5°C.

111.4.3.1.3 Mixing

The tank containing the aggregates to be mixed is placed on the mixer, trying to keep temperature losses to a minimum. The mixer is run for 30s ± 5s to homogenise the sands. If the binder is poured in once, the mass of binder must not exceed 1% of the theoretical binder mass, otherwise the batch will be rejected. The mixing time must result in a visually homogeneous mix, so the total mixing time is between 2 and 3 minutes. Once mixing is complete, the mixture must be used immediately before cooling, otherwise the paste will be rejected. The mass of the bitumen is calculated from the mass of the aggregates as shown in the following formula (Eq. III.9):

$$ML = \frac{(MA * TL)}{100} \qquad (g) \qquad (Eq.\ III.9)$$

Or

ML: mass of bitumen (binder) used (g) ;

MA: the mass of the mixture of aggregates used (g) ;

TL: the bitumen (binder) content used in a mix (%).

111.4.3.2 Making the test tubes for the "Marshall" test

III.4.3.2.1 Mould filling and compaction

Weigh a quantity m equal to 1200g of mixture to within 0.1% in relative value. After placing a paper disc at the bottom of the mould, the moulds were heated to the reference temperature for preparing the test tubes for a minimum of 2 hours, the riser was put in place and the mixture was introduced into the mould in a single operation, lightly coated with glycerine sodium oleate (Figure III.8).

FigureIII.8 - Filling the mould and compacting.

The second paper disc is then placed on top of the mixture. Place the moulds in the machine and compact with 50 strokes for 55 ± 5 seconds. The mould is disassembled and reassembled, swapping the base and the top, and the grooming is repeated. The total number of strokes is 100. The mould is kept for at least 5 hours at room temperature (15 to 25° C) after compacting.

III.4.3.2.2 Demoulding

Once the moulds have cooled and are ready to be demoulded, the test tube is passed through the riser with the aid of the extractor piston and the press, as shown in (Figure III.9). The other test-tubes are made by following the same steps above.

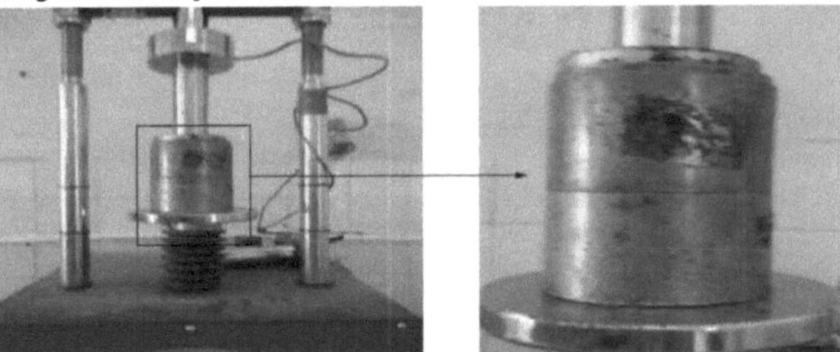

Figure III. 9 - Demoulding operation.

When the test specimens are ready, the Marshall test is started and the following parameters are determined:

The apparent density of the test tube and the theoretical (absolute) density of the test tube ;

Compactness of the test tube ;

% of residual voids or voids in the test piece; % of voids occupied by air and bitumen (void in the aggregate) and % of voids filled by bitumen;

Marshall deformation or creep in *"mm"* and Marshall stability in *"Kg"*.

III.4.3.2.1 Test tube after preparation

As with the other analysis methods, the Marshall test tube is determined by making a cylindrical test tube (Figure III.10), with an indicative mass of 1200g, a diameter of 105 mm and a theoretical height of 63.5 mm.

Figure III.10 - Specimen after preparation for the Marshall test.

1.1.1.1.1 Production of test tubes for the "Duriez" test

1.1.1.1.2 Mould filling and compaction

A piston is placed at the bottom of the mould. The mixture is introduced in one go into the mould, which is very lightly coated with glycerine sodium oleate (the moulds are brought to the reference temperature for preparing the test tubes for at least 2 h) before the operation (Figure III.11).

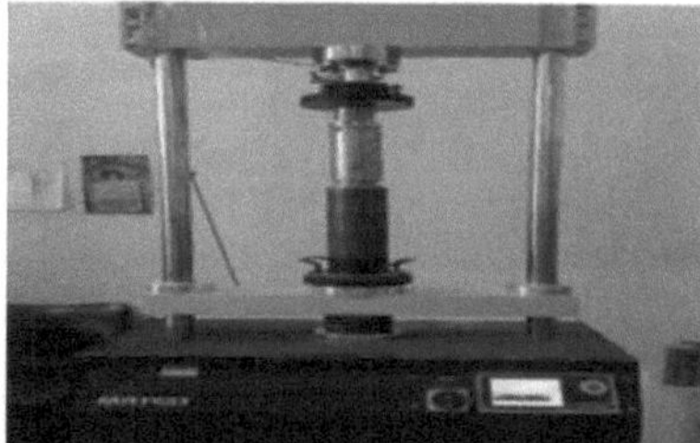

Figure III. 11 - Filling the mould and compacting.

The full moulds are then placed in an oven at a temperature close to the reference temperature, where they are left for between 30 minutes and 2 hours. The full cylinder is adjusted and carried between the press platens.

The compaction of the specimens must be carried out by double effect (the operations must be carried out in such a way as to avoid temperature losses as much as possible). The pressure is maintained for five minutes, bearing in mind that the load applied is of the order of 60 kN ± 0.5%. The test specimens are kept lying in their moulds for at least 4 hours until they reach room temperature, then they are demoulded.

1.1.1.1.3 Demoulding

Demoulding is carried out using a press. The extracted specimens are divided into two batches, the first batch specimens (2 in number) are intended to determine the apparent density and the second batch specimens are intended for the compressive strength test.

The test tubes from the second batch were divided as follows: two test tubes were made without immersion and the other two were made with immersion, all of which were placed in a special cabinet at 18°C ± 0.5°C for 7 days. The parameters obtained from this test are :

The apparent density of the test tube ;

The true or theoretical (absolute) density of the test tube ;

Compactness of the test tube ;

%   residual voids or voids in the test tube ;

%   roof spaces with bitumen;

Compressive strength $"r"$ after immersion for 7 days at 18°C in Kg/cm^2 ;

Compressive strength $"R"$ before immersion at 18°C in Kg/cm^2 ;

The $"r/R"$ ratio ;

The $"W"$ soaking percentage.

1.1.1   .3.3.1 Test tube after preparation

The *"Duriez"* test tube is determined by making a cylindrical test tube (Figure III.12), with an indicative mass of 1000 g, 80 mm in diameter and 190 mm high.

Figure III.12 - Specimen after preparation for Duriez test.

III.5 Performance verification tests on formulated asphalt mixes

There are several types of tests to determine the performance of formulated mixes:

1.1.2   Rutting test

The rutting test is used to study asphalt mixes for roads with heavy and very heavy traffic. It is used to assess the rutting resistance of wearing courses and base courses intended for the types of traffic described above, under conditions comparable to those encountered on carriageways. The test is characterised by the determination of the rutting depth caused by the repeated passage of a tyre over a slab of coating at 60°C for surface courses and at 50°C for base courses.

The $"LPC"$ rutting tester from the Laboratoire Central des Ponts et Chaussees (LCPC) in France has been used successfully for more than 20 years to prevent the risk of rutting of pavements, particularly in France, Belgium and Switzerland (LAVOC). It is also used by laboratories in Algeria (Figure III.13).

The *LPC* orniereur is a laboratory instrument for testing :

The susceptibility of asphalt mixes to rutting;

The resistance of bituminous wearing courses, particularly surface dressings and thin overlays, to tangential stresses.

Figure III.13 - *LCP rounder.*

The parallelepipedic test tube is made on an *"LPC"* compaction table or taken from a roadway. The rounder, equipped with 2 independent cradles, enables 2 test tubes to be tested in parallel, with or without identical parameters. The surface of each specimen is subjected to the stresses of a wheel mounted on a carriage driven by a sinusoidal reciprocating motion. The load is applied by a cylinder acting on the support cradle of each test piece. A lateral shearing effect can be introduced by changing the angle of the wheel in relation to the direction of travel. The simulator is enclosed and thermally insulated. This means that the tests can be carried out at a regulated and constant temperature. The main characteristics of the *"LCP"* rut are as follows:

Plate dimensions: L * B * H = 500 * 180 * max. 140 mm ;

Cylinder load: max. 5 kN ;

Tyre pressure: 0.100 to 0.700 MPa ;

Frequency: 7200 passages/hour;

Test carried out simultaneously on 2 plates, test time for one formula: 5 days.

The result of the test is the rut depth or deformation as a percentage of the initial height of the test piece Pi%. The permissible deformation *"Pi%"* is well below 10%.

1.1.3   Fatigue test

This test consists of subjecting a trapezoidal test specimen of concrete embedded at its base to bending through its free edge. This load is imposed by displacement. In other words, a sinusoidal displacement of constant amplitude is imposed at the end of the specimen and it is assumed that failure is reached when the force required to obtain the deformation is equal to half the initial force.

The two-point bending fatigue test consists of stressing a trapezoidal specimen embedded in its large base in bending. This load can be applied in two ways:

Imposed displacement: a sinusoidal displacement of constant amplitude is imposed on the end of the test piece and it is assumed that failure is reached when the force required to obtain the deformation is equal to half the initial force;

Imposed force: a sinusoidal force of constant amplitude is applied to the end of the test piece and it is assumed that failure is reached when the modulus of rigidity is half that of the initial modulus.

At one temperature, the test is repeated at several strain levels, enabling the fatigue curve *"N=f(e)"* to be established, where *"N"* is the number of loads causing failure and *"e"* is the strain on an outer fibre of the specimen (h = 250, B = 56, b = 25, e = 25 mm). It should be noted that the excitation frequency is 10 to 40 Hz and the displacement amplitude at the head of the specimen is 0 to 1.4 mm.

1.1.4   Complex modulus test

This test characterises the viscoelastic behaviour of mixes as a function of frequency and temperature. The modulus test is carried out on a trapezoidal test specimen embedded at its base and on the free end, a very small sinusoidal displacement of constant amplitude is imposed, creating a bending of the

64

test body by simulating the effect of traffic. From the resulting force, the modulus is calculated over a temperature range from *"-10 to 40°C"*, and for each temperature, four frequency levels are used: *"1, 3, 10 and 30 Hz"*.

III.6 Conclusion

This final chapter describes most of the different types of bitumen and bituminous concrete, the various mix design methods and the main identification tests on the materials used in bituminous concrete mix design, as well as the various performance verification tests on formulated mixes.

GENERAL CONCLUSION

In conclusion, this book entitled **"Road Pavement Degradation, Road Testing and Formulation"** provides an in-depth exploration of the main aspects relating to the construction and re-evaluation of road pavements. The three chapters cover key areas of road engineering, providing readers with a comprehensive understanding of the processes and associated challenges.

In the first chapter, we examined the criteria, modes and methods for assessing road pavement deterioration. The second chapter focused on road identification tests, enabling us to identify the characteristics and properties of road pavements. We examined both laboratory and field test methods, taking into account traffic loads and environmental conditions. These elements are essential for the appropriate design of road pavements. The third chapter immerses us in the formulation of bituminous concrete used in pavement construction. We explore the components, mixing and manufacturing processes, as well as pavement durability and performance considerations. Understanding the formulation of materials is crucial to ensuring high quality pavements that resist degradation.

This book has been designed to conform to the official syllabus of the Algerian Ministry of Higher Education and Scientific Research (MESRS). It aims to help students of road engineering to develop the skills needed to meet the challenges of road pavement construction.

Offering a combination of theoretical and practical knowledge, each chapter provides an in-depth understanding of the different stages in the construction, evaluation and design of asphalt mixes for road pavements.

This book is an invaluable guide for students of road engineering, helping them to develop sound expertise in the construction and evaluation of road pavements. It will enable them to successfully meet the challenges they will face in their future professional practice, contributing to the creation of high quality, sustainable road infrastructure.

BIBLIOGRAPHICAL REFERENCES

1. **Alloul, B. (1981)**. Etude geologique et geotechnique des tufs calcaires et gypseux en vue de leur valorisation en technique routiere. Doctoral thesis, University of Paris IV.

2. **Asphalt Institute (1997)**. Mix Design Methods. Manual Series No.2 (MS-02). Asphalt Institute, Lexington, KY.

3. **Asphalt Institute (2001)**. Superpave Mix Design. Superpave Series No.2 (SP-02). Asphalt Institute, Lexington, KY.

4. **ASTM D4318 (2000)**. Standard test methods for liquid limit, plastic limit, and plasticity index of soils. Annual book of ASTM Standards, American Society of Testing and Materials, USA, 04:08, doi: 10.1520/D4318.

5. **ASTM D698 (2000)**. Standard test methods for laboratory compaction characteristics of soil using standard effort. Annual book of ASTM Standards, American Society of Testing and Materials, USA, 04:08.

6. **Atterberg, A. (1911)**. The behaviour of clays with water, their limits of plasticity and their degrees of plasticity. Internationale Mitteilungen fur Bodenkunde, vol. 1, p. 10-43.

7. **BLPC (1998)**. Bulletin des laboratoires des Ponts et Chaussees.

8. **Bonaquist R.F., Christensen D.W., Stump W. (2003)**. Simple performance tester for Superpave mix design: first article development and evaluation. NCHRP report 513, Washington D.C.

9. **CEBTP-LCPC (1985)**. Manuel pour le renforcement des chaussees souples en pays tropicaux. Republique Fran^aise, Ministere des relations exterieures cooperation et developpement, 166p.

10. **Centre de Recherches Routieres, CRR (1987)**. Code de bonne pratique pour la formulation des enrobes bitumineux denses. Recommendations C.R.R. - R 61/87.

11. **Centre de Recherches Routieres, CRR (1997)**. Code de bonne pratique pour la formulation des enrobes bitumineux denses. Recommendations C.R.R. - R 69/97.

12. **Coquand, R. (1969)**. Routes-Vol. 1: Circulation-trace-construction; Vol. 2: Construction et entretien. Paris: Eyrolles. Pp. 285.

13. **Costet, J and Sanglerat, G. (1983)**. Cours pratique de mecanique des sols.4eme trimestre: Dunod, Pp. 442.

14. **CTTP (1995)**. Guide de l'entretient routier. Controle technique des travaux publics, Algerie.

15. **CTTP (1996)**. Guide de rehabilitation des routes. Controle technique des travaux publics, Fascicule 01, Algerie.

16. **CTTP (2000)**. Recommandation algerienne sur l'utilisation des bitumes et enrobes bitumineux a chaud.

17. **EN 933-2 (2020)**. Tests for geometrical properties of aggregates - Part 2: Determination of particle size distribution - Test sieves, nominal size of apertures.

18. **EN 933-8 (1999)**. Tests for geometrical properties of aggregates - Part 8: Assessment of fines - Sand equivalent test, Directive 89/106/EEC, technical corpus CEN/TC. p. 154.

19. **Fascicule 3 (2015)**. Catalogue de dimensionnement des Chaussees Neuves Fascicule3 Fiches Techniques de Dimensionnement. Pp. 68-73.

20. **Gadouri, H, Harichane, K, and Ghrici, M. (2019)**. Effect of sulphates and curing period on stress-strain curves and failure modes of soil-lime-natural pozzolana mixtures. Marine Georesources & Geotechnology, 37(9), 1130-1148.

21. **Joeffroy, G. and Sauterey, R. (1991)**. Dimensionnement des chaussees. Paris: Presses de l'Ecole Nationale des Ponts et Chaussees, Pp. 173-174.

22. **Joubert, P et al (2006)**. Application du modele GiRR pour la programmation de travaux d'entretien au Montenegro. Laboratoire central des ponts et chaussees (LCPC), Bulletin des laboratoires des Ponts et Chaussees, (BLPC). n°265, Pp. 61-87.

23. **LCPC (1991)**. La methode VIZIR : Methode assistee par ordinateur pour l'estimation des besoins en entretien d'un reseau routier. Laboratoire Central des Ponts et Chaussees (LPCP), Paris, Pp. 63.

24. **LCPC-GTR (2000)**. Realisation des remblais et de couches de forme. Fascicule 1, Principes generaux, Cerema, France.

25. **LCPC-IFSTTAR (1998)**. Catalogue des degradations de surface des chaussees. Institut Fran^ais des sciences et techniques des reseaux, de l'Amenagement et des transports (IFSTTAR). Laboratoire centrale des ponts et chaussees (LCPC), n° 58, Boulevard Lefebvre-75732, Paris, CEDEX 15.

26. **LCPC-SETRA (1985)**. Directive pour la realisation des assises de chaussees en sables traites aux liants hydrauliques. Paris: Bagneux, France.

27. **LCPC-SETRA (1998).** Catalogue des structures types de chaussees neuves. Paris: ministres de l'equipement des transports et du logement; Bagneux, Pp. 297.

28. **LCPC-SETRA (2000).** Guide des terrassements routiers : Realisation des remblais et des couches de forme. Guide technique, France.

29. **LCPC-VIZIR (1991).** Methode assistee par ordinateur pour l'estimation des besoins en entretien d'un reseau routier, 64p.

30. **MTQ-AIMQ (2002).** Manuel d'identification des degradations des chaussees souples (MIDCS). Ministere des transports de Quebec (MTQ), Association des ingenieurs municipaux du Quebec (AIMQ), Pp. 58, Quebec, Canada.

31. **NF EN 933-8 (1999).** Essais pour determiner les caracteristiques geometriques des granulats - Partie 8: Evaluation des fines - Equivalent de sable, Paris, Association Fran^aise de Normalisation (AFNOR). replaces the French experimental standards P. 18-597, P. 18-598.

32. **NF P 15-301 (2015).** Hydraulic binders - Definition - Classification and specifications of cements.

33. **NF P 94-093 (2014).** Sols : reconnaissance et essais - Determination des caracteristiques de compactage d'un sol par l'essai Proctor normal et Proctor modifie.

34. **NF P 98-253-1 (1991).** Deformation permanente des melanges hydrocarbones, partie 1 : Essai d'ornierage. p. 11, France.

35. **NF P94-051 (1993).** Sols : Reconnaissance et essais - Determination des limites d'Atterberg - Limite de liquidite a la coupelle - Limite de plasticite au rouleau.

36. **Swiss standard VSS SN 640 925a (1997).** Survey and evaluation of road conditions. Switzerland.

37. **NTAR-B40 (1977).** Normes techniques d'amenagement des routes (NTAR): Etudes generates techniques et economiques des amenagements routiers. Niveau de service et normes, Ministre des Travaux Publics, Algerie. https://geniecivilettravauxpublics.blogspot.com/2012/09/normes-techniques-algerienne-b-40.html.

38. **Perret J., Dumont A.-G., Turtschy J.-C., Ould-Henia M., (2001).** Evaluation des performances de nouveaux revetements : 1 partie : enrobes a haut module. OFROU Report No. 1000.

39. **WesTrack Forensic Team Consensus Report "WTFTCR" (2001).** Superpave mixture design guide. Washington D.C.

40. **XP P 18-540 (1997).** Aggregates: Definitions, conformity and specifications. AFNOR, P 18-540, ICS:91.100.20, France.

I **want** morebooks!

Buy your books fast and straightforward online - at one of world's fastest growing online book stores! Environmentally sound due to Print-on-Demand technologies.

Buy your books online at
www.morebooks.shop

Kaufen Sie Ihre Bücher schnell und unkompliziert online – auf einer der am schnellsten wachsenden Buchhandelsplattformen weltweit! Dank Print-On-Demand umwelt- und ressourcenschonend produzi ert.

Bücher schneller online kaufen
www.morebooks.shop

Printed by Books on Demand GmbH, Norderstedt / Germany